Enhancing Change through Science Centres

Enhancing Change through Science Centres

— Editors —

Aman Kumar Maulloo
Lebs Mphahlele

CENTRE FOR SCIENCE & TECHNOLOGY OF THE NON-ALIGNED AND OTHER DEVELOPING COUNTRIES
(NAM S&T CENTRE)

2010
DAYA PUBLISHING HOUSE
Delhi - 110 035

ISBN 81-7035-626-1
ISBN 978-81-7035-626-4

Centre for Science and Technology of the Non-Aligned and Other Developing Countries (NAM S&T Centre)
Core-6A, 2nd Floor, India Habitat Centre, Lodhi Road,
New Delhi-110 003 (India)
Phone: +91-11-24644974, 24645134, Fax: +91-11-24644973
E-mail: namstct@gmail.com
Website: www.namstct.org

Published by : **Daya Publishing House**
1123/74, Deva Ram Park, Tri Nagar, Delhi - 110 035
Phone: 27383999, Fax: (011) 23260116
e-mail: dayabooks@vsnl.com
website: www.dayabooks.com
Showroom : 4760-61/23, Ansari Road, Darya Ganj
New Delhi - 110 002
Phone: 23245578, 23244987

Laser Typesetting : **Classic Computer Services**
Delhi - 110 035

Printed at : **Chawla Offset Printers**
Delhi - 110 052

Printed in India

Foreword

The year 2006 marked an important step in the implementation of the National Norms and Standards for a Network of Science Centres in South Africa. This was when we adopted our National Roll-out Plan for the Establishment of a Network of Science Centres in South Africa. Among other things, the plan prioritises networking between science centre communities, particularly intergovernmental networking.

South Africa's co-organisation and hosting of the 5th International Workshop on Enhancing Change through Science Centres was a milestone in the intergovernmental networking led by my department, and I would like to thank Prof. Arun, Director of the Non-Aligned Movement Science and Technology Centre, for giving us this opportunity. I would also like to express my gratitude to the delegates, the Southern African Association for Science and Technology Centres (SAASTEC), the Sci-Bono Discovery Centre and my staff, without whom the workshop would not have been such a success.

Among the most valuable outcomes of the 5th International Workshop was the sharing of information on how to use easily accessible and cheap materials to develop interactive science and technology exhibits, which are an integral part of increasing the participation and performance of our young people in these fields.

One of the highlights of the workshop was the participation of the five members of the International Programme Committee of the Science Centre World Congress (SCWC), who were instrumental in the development of the Johannesburg Declaration.

The 5th International Workshop's recognition of the fact that some parts of past declarations had not been implemented marks it as a turnaround point in the history of science centres in developing countries.

For South Africa, the co-organisation and hosting of the 5th International Workshop on Enhancing Change through Science Centres was an important rehearsal as the country gears to host the 6th Science Centre World Congress in Cape Town in

2011. The South African Chapter of SAASTEC will be organizing and hosting the 6th SCWC in collaboration with my department and the MTN ScienCentre in Cape Town.

As the first African country to organize and host the SCWC, our intention is to use it as a platform for both developed and developing countries to share ideas about the role science centres can play in a developing society.

I hope to see all the delegates who attended the 5th International Workshop on Enhancing Change through Science Centres at the 6th Science Centre World Congress in 2011.

H.E. Mr. M Mangena

Minister of Science and Technology, South Africa

Preface

This book is a collection of 28 invited papers and keynote addresses written by some of the participants at the 5th international workshop on 'Enhancing Change through Science Centres' held in Johannesburg, South Africa, from 25-28 February 2008 organized by the Centre for Science and Technology of the Non-Aligned and Other Developing Countries (NAM S&T Centre) in collaboration with the Department of Science and Technology (DST), Republic of South Africa. This workshop gathered together the specialists and professionals from the government and private institutions and departments in the developing countries with relevant background and active involvement in development policy and implementation of S&T programmes through science centres and museums.

The geographical distribution of authors is extensive with the majority coming from Asia. More interestingly there has been a growing interest from African and South American regions in the last few years. It is noteworthy that most papers managed to focus on the theme of the workshop, namely, "Enhancing Change through Science Centres".

The selected papers have been classified depending on their affinities and grouped under four main parts of the book. The keynote addresses presented in the first part of the book are from two invited speakers with long dedicated careers to Science Centres and coming from two extreme geographical locations. The first presentation on "The Science Centre Movement in India" is by Mr. Ingit K. Mukhopadhyay, former Director General of the National Council of Science Museums (NCSM), India, who has served for 31 years at the NCSM and who has recently been awarded the prestigious 2008 Roy Shafer Leading Edge award by ASTC in the Experienced Leadership category. It should be highlighted that it is the first time that this award has come to Asia. The second presentation is on "The Role of Science and Technology in Developing Countries" written by Mr. Joaquin Fargas, from Argentina. Mr. Fargas founded the 'Centro Cientifico Tecnológico Interactivo (CCTI) in 1990 and is the Director of Arts, Science and Technology Global Museum. He was awarded the Red-Pop/UNESCO 2005 award for the Popularisation of Science and Technology.

The second part of the book is entitled "Evolution of Science Centres" and is a collection of nine papers from eight different countries. Each article deals with a different topic relevant to the current trend of Science Centres. The topics covered include: social appropriation, digitalisation, outreach policy, religion and history, bilateral collaboration, awareness campaigns, science education, and the role of a science city and specialised centres. All the major themes are of current interest and are well presented.

The third part of the book is termed as "Rationale for Establishment of Science Centres". This section comprises of 11 papers from 10 participating countries. These articles highlight the growing interest and importance for setting up science centres to dynamically adapt to each country's needs. The concepts developed include volunteer programmes, performance in schools, investment in S&T, sustainable development, science parks and science culture.

The last part of this publication entitled "Science Centre Policies" presents a collection of six articles on International support, research and networking policies. Although these contributions are not directly linked with Science centres, they do indirectly enhance the promotion of Science and Technology worldwide.

The overall editing process was very enriching and constructive. Following the inputs from authors, a framework was developed and adhered to ensure timely output. Revision was then requested from concerned authors and the NAM S&T Centre did the fantastic monitoring and finishing job for this book to be available.

It was observed that there were few contributions on important 21st century topics such as innovation, science exhibit design, indigenous knowledge systems, 'science parks' and culture. As agents of change, science centres can and should play a role in interrogating the role of science in promoting this knowledge category. We expect to have more participation in this dimension in future workshops.

We hope that this achievement will meet the expectations of those involved in Science Centres and also those who are keen to set up such institutions in their locations. We also wish that the foundation laid thus far will spur more publications in this domain.

We express our sincere gratitude to Professor Arun Kulshrestha and the NAM S&T Centre for having given us this opportunity to edit this second volume of workshop proceedings. Our knowledge about science centres in relation to countries' needs have definitely grown.

Aman Kumar Maulloo

Lebs Mphahlele

Introduction

In recent years we have been witnessing unprecedented growth in various fields of science and technology, such as nanotechnology, information and communications technology, genomics, genetic engineering and space science, which have thoroughly influenced or are about to transform our lives. In addition, in our rapidly changing world, there is a great need for the public understanding of science and technology to create more science literate society at all age levels. It is an undisputed fact that every community needs to complement their formal education system with more experimental and enjoyable learning. Science Centres provide one unique and critical response to these needs.

Science is a part of our culture and society, and science centres and museums provide a distinct network between the community, the school and the home, giving them a unique niche in the educational infrastructure. The public in the developing countries is well educated, emancipated and demanding and expects authenticity, scientific integrity and social relevance. Planning a viable science centre or museum therefore requires an appreciation of the historical past, as well as an analysis of the environment in which it is supposed to operate, including its public and its stakeholders.

Science centres and science museums are vital and powerful media for popularization of science and creating a scientific temper across the globe. Many developing countries may not comprehend the significance of this mode of science communication because of political and other problems, but the time has come to educate the populace through sharing of knowledge and information and transfer of technology. Therefore the Governing Council of the NAM S&T Centre has accorded high priority to popularize the concept of science centres and science museums among developing countries. In this context, the NAM S&T Centre has successfully organized a series of international and regional level workshops and conferences in different countries, namely, in India (Kolkata) in partnership with National Council of Science Museums (NCSM) in November-December 2002; in Colombia (Bogota) in association with Andrés Bello Agreement, COLCIENCIAS and MALOKA Science Centre in March 2004; in Vietnam (Hanoi) in association with the Centre for Regional Research and Development (CRD) under the Ministry of Science and Technology (MOST) of Vietnam in October 2004; and in Zambia (Lusaka) jointly with their National Science Centre in January 2006.

In order to address some of the pertinent issues and review the latest status of development related to science centres, science museums, science popularization and science education, the Centre for Science and Technology of the Non-Aligned and other Developing Countries (NAM S&T Centre) organized its 5th international workshop in the series titled 'Enhancing Change through Science Centres' during 25-28 February 2008 at Johannesburg, South Africa jointly with the Department of Science and Technology (DST), Government of South Africa. This workshop was intended as a forum for the scientists, technologists, innovators, industry representatives and policy makers to brainstorm on the importance and functionality of science centres, an issue of vital significance for the sustainable growth of science in their countries.

The Workshop was attended by more than 45 heads and senior experts on science centres and science museums and policy makers from 22 Countries. The overseas participants were from Argentina, Colombia, Cuba, Ghana, India, Indonesia, Iran, Lesotho, Malawi, Malaysia, Mauritius, Namibia, Nigeria, Pakistan, Serbia, Sri Lanka, Tanzania, Turkey, Vietnam, Zambia and Zimbabwe. The overall technical programme of the Workshop was coordinated by Mr. Isaac Ramovha from the Science and Youth Unit of DST jointly with me.

The six technical sessions of the Workshop were respectively chaired and co-chaired by Dr. I.K. Mukhopadhyay (India) and Dr. Derek Fish (Unizul Science Centre, South Africa); Dr. E. Hoyos (Colombia) and Mr. Alfred Tsipa (South African Agency for Science and Technology Advancement (SAASTA); Mr. Nakano Hideo (Ghana/ Japan) and Ms. Rudy Hurak (Manager, Sci-Enza, Pretoria); Ms. Tengku Nasariah Ibrahim (Malaysia) and Ms. Julie Claverdon (Sciencentre, Cape Town); Mr. J. Fargas (Argentina) and Dr. Rufus Wesi (Boitjhorisong Resource Centre, Sasolburg); and Dr. A. K. Maulloo (Mauritius) and Dr. Michael Peter (Chief Operating Officer, Sci-Bono, Johannesburg). All the overseas speakers presented the status of the science centres, science museums and science popularization related efforts in their respective countries. Mr. Nakano Hideo, JICA Senior Volunteer from Ghana gave a talk with demonstration of several inexpensive science education materials. The speakers from South Africa were Mr. K. M. Lebs Mphahlele, Manager (Science and Youth), DST, and Mr. Alfred Tsipa of SAASTA. Dr. Derek Fish of Unizul Science Centre and Dr. Stuart Hopwood, Manager: Exhibitory, Sci-Bono made a highly entertaining joint presentation. Besides the technical programme the participants made study visits to the Origins Centre, Sci-Bono Discovery Centre, Sci-Enza Science Centre and Sterkfontein Caves.

The present publication comprises a compilation of 28 scientific articles and review papers. I would like to express my deepest gratitude to His Excellency Mr. Mosibudi Mangena, Honorable Minister of Science and Technology, South Africa for his personal patronage in all phases of the organization of the workshop right from conceiving this event till writing a 'Foreword' for this publication. I am also thankful to Dr. Phil Mjwara, Director General of Science and Technology, South Africa; Ms. Lindiwe FLNW Lusenga, General Manger for Africa and Multilateral Cooperation, Department of Science and Technology; Mr. Lebs Mphahlele and Mr. Isaac Ramovha, Department of Science and Technology for their comprehensive support and guidance in this successful event.

Last, but not the least, I would also like to acknowledge the dynamic involvement and untiring efforts of Dr. Aman Kumar Maulloo, Director, Rajiv Gandhi Science Centre of Mauritius and Mr. K. M. Lebs Mphahlele, Manager (Science and Youth), Department of Science and Technology of South Africa for technical editing of this publication and suggesting the future course of action while reviewing the contents. The valuable services provided by the entire team of the NAM S&T Centre, particularly by Mr. M. Bandyopadhyay, Dr. V. P. Kharbanda, Mrs. Sudipta Chakraborty and Mr. Pankaj Buttan in compiling the presented papers and giving a shape to this volume are also deeply appreciated.

I hope that this publication would undoubtedly be a valuable reference material for the entire scientific community especially those working in the field of science centres and museums, and will help the developing countries in executing proper plans and strategies for inspiring the scientific values and temper among their people.

Arun P. Kulshreshtha

Director, NAM S&T Centre

Contents

Part III: Rationale for the Establishment of Science Centres

Part IV: Science Centre Policies

Part I
Keynote Address

Chapter 1

The Science Centre Movement in India

Ingit K. Mukhopadhyay

National Council of Science Museums, Kolkata, India

E-mail: ncsmin@vsnl.com, ncsmin@giascl01.vsnl.net.in

ABSTRACT

In the early 70s, when India had already passed through three Five-Year Plans, science education started getting into the national priority list. The Planning Commission of India at this time set up a Task Force to determine how science could be popularized further through science museums in India. The Task Force, after elaborate examination of the issue recommended a course of action to be taken for proper use and growth of these institutions. It saw great potential in the science museums for creating awareness among the general public on scientific and technological issues and for creating a scientific temper among the people at large. It is true that in India at present a career in engineering or information technology is highly rewarding and a career in science is by far less attractive in terms of remuneration, career growth and other prospects. Although this is the major reason for a decline in career in science in India, at present there are other reasons related to the formal undergraduate science education programme which lacks excitement of learning in science. Science Centre activities can very effectively and substantially fill up what is lacking in the formal science education system and possibly bring back the thrill in science learning through an interactive science approach.

Keywords: *India, Science centres, Science museums, Science popularization, scientific temper, Technology developments, Science movement.*

1. Introduction

The first science museums in India were set up in the year 1953–one inside the campus of Birla Institute of Technology and Science (BITS) in Pilani and the other one inside the premises of the National Physical Laboratory (NPL) at New Delhi. The first one had an objective of exposing various principles of technology and science to the students studying in the University and also the local students. The Museum at New Delhi was to display the post-independence developments in science and technology in India. The museum at New Delhi wound up after a short tenure due to various reasons but the Birla Museum is still operating from the same venue. The Government of India, however, considered non-formal science education, an effective tool to supplement formal science education to be important and in 1956 an organized effort under the aegis of the Government of India initiated the work for setting up the first science museum in India for taking the messages of Science and Technology to the people at large. The objectives, however, had a strong component for showing post-independence Indian achievements in Science and Technology but in general, the concept followed the European model in which the history of development of a particular area of Science and Technology up to the contemporary state was told through artifacts, working models and other media. In 1959, the Birla Industrial and Technological Museum (BITM) were set up in the southern part of Kolkata under the governance of the Council of Scientific and Industrial Research (CSIR), an autonomous organisation under the Government of India. The exhibit galleries and activities were designed following the European museums particularly the Deutsches Museum in Munich and the Science Museum in London. The BITM opened its doors with 7 galleries on Metallurgy of Iron, Steel and Copper; Petroleum; Electricity; Nuclear Physics; Electronics; Television and Optics. Soon the new Museum became the talk-of-the-town and true to the fact that the city of Kolkata had been the cradle of science education in the country from the time under the British Rulers, the students took immense interest in visiting BITM. The schools in and around Kolkata organized visits regularly with an intention to see working exhibits and watch science demonstration lectures. As the popularity of BITM grew, a demand for another scientific and industrial museum in Bangalore was raised. The City of Bangalore in the early 60s, was growing to become the hub of scientific activities in India with new research institutions established there and industries with modern technologies emerging in and around the city. The Visvesvaraya Industrial and Technological Museum (VITM) started operation in 1965 and as expected, became very popular not only in Bangalore but also in the entire south India.

In the early 70s, when India had already passed through three Five-Year Plans, science education started getting into the national priority list. The Planning Commission of India at this time set up a Task Force to determine how science could be popularized further through science museums in India. The Task Force, after elaborate examination of the issue recommended a course of action to be taken for proper use and growth of these institutions. It saw great potential in the science museums for creating awareness among the general public on scientific and technological issues and for creating a scientific temper among the people at large. Some of the major recommendations of the Task Force were:

1. Science Museums are to be developed in three levels–national, regional and district level supported by Mobile Science Exhibition units;
2. The Science Museums to be set up in different parts of the country need not be of the same type in order to follow a uniform model in terms of their size, nature, scope and content. However, the museums may have a common basic core element supplemented by sections relevant to the local conditions; and
3. For the planning of the network of science museums, there should be a separate executive agency with required specialization and expertise.

In the mean time, some pilot projects for setting up small science museums in district towns were taken by BITM in eastern India and VITM in southern India. These projects received wide spread response. The Task Force recommendation in the later days for creation of district level science centres supported the small town science museums concept. Another important development in the late 60s was the commencement of Mobile Science Exhibitions. These exhibitions set up in large buses called Museobuses, were sent to deep rural areas for taking the message of Science and Technology to the village dwellers who for various reasons could not travel to the nearby towns or large cities to visit science museums. In fact, the opportunity of having an interface with science and technology issues and products, were virtually absent for them. The Mobile Science Exhibition activity merged comfortably with the national priority for spreading education in the rural areas. India is a vast country with majority of its people living in villages and therefore, the scope of spreading the message of Science and Technology through Mobile Science Exhibition units had immense potential to succeed.

At this time, another major science museum in another large metropolitan city of Mumbai (Bombay) was being planned. However, the world of science museums at this point of time was influenced by the interactive exhibits introduced in places like Exploratorium. The Mumbai set up, therefore, was being planned to be the first interactive science centre deviating from the old path of artifact-based science museums.

On the basis of the recommendation of the Task Force made by the Planning Commission, the Government of India decided to set up a new autonomous society under the title National Council of Science Museums, de-linking it from the CSIR. This move paid very good dividend as the science museums under the CSIR were already facing a major priority crisis, the primary area of activities of CSIR being scientific and industrial research and not science communication. The National Council of Science Museums on the other hand, was given autonomy to govern the existing science museums and to develop new science centres as per the recommendation of the Task Force. During the next three decades, the growth of science centres in India had been meteoric.

This trend is clearly depicted in the Figure 1.1.

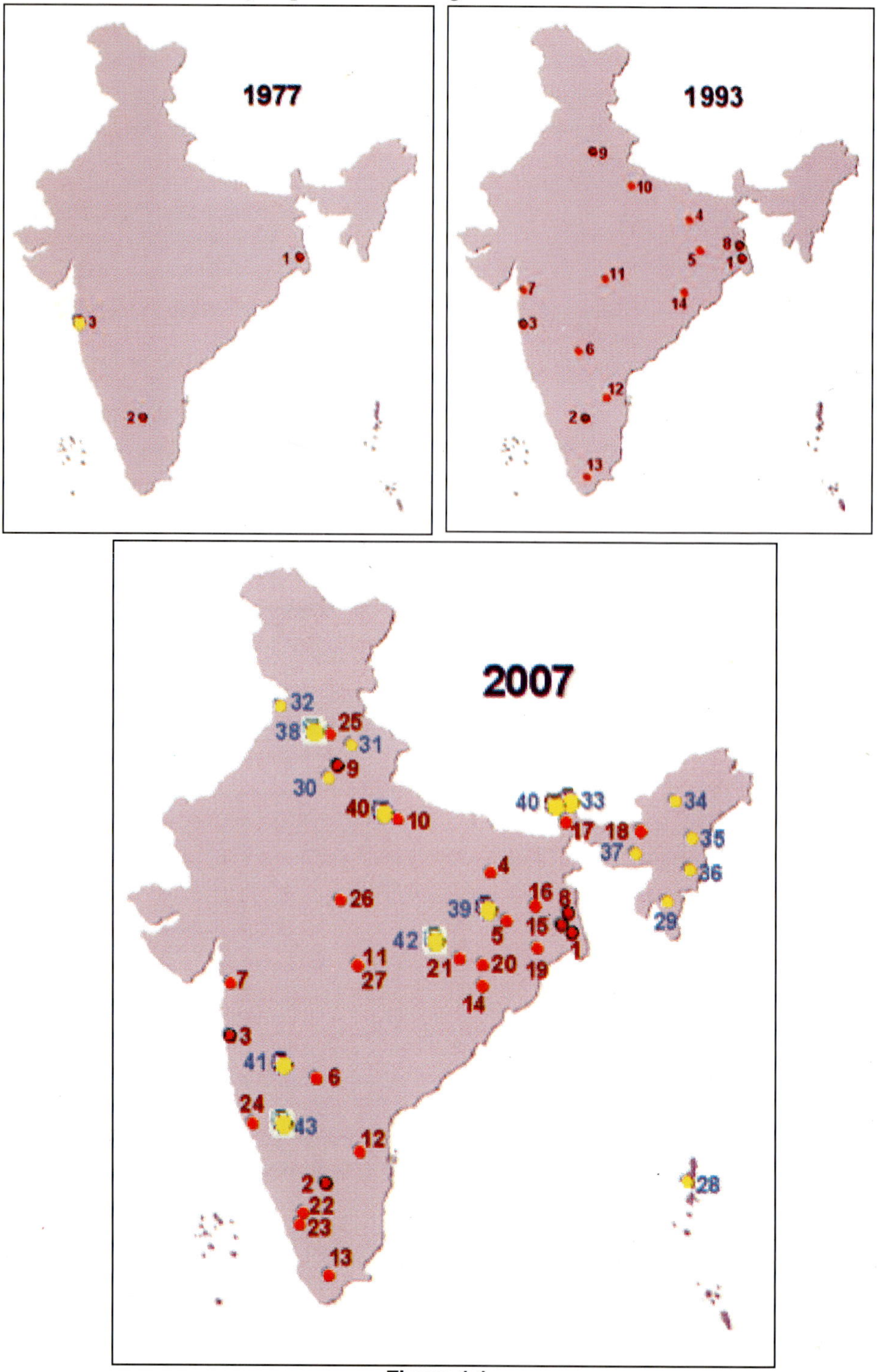

Figure 1.1

MAJOR CENTRES OF NCSM AT KOLKATA, BANGALORE, MUMBAI AND DELHI

Birla Industrial and Technological Museum, Kolkata

Visvesvarya Industrial and Technological Museum, Bangalore

Nehru Science Centre, Mumbai

National Science Centre, Delhi

Science City, Kolkata

Science City, Kolkata

NCSM Hqrs. and Central Research and Training Laboratory, Kolkata

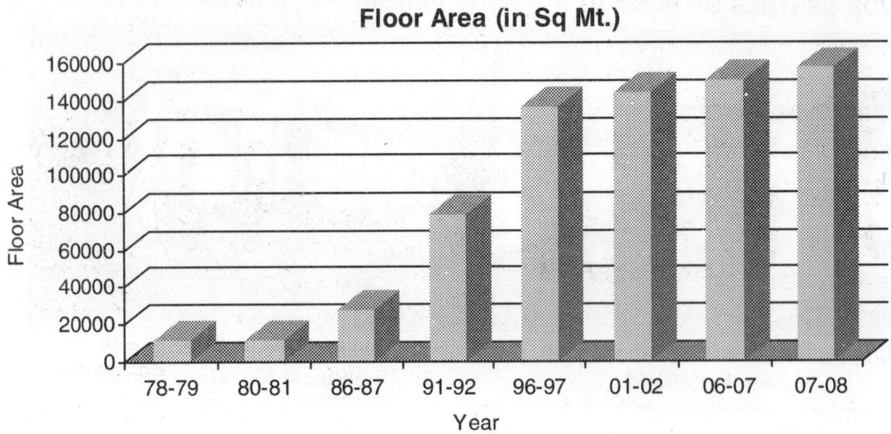

2.0 Priorities and Policies

Infrastructure development had been NCSM's top priority in 1978, when the Council started functioning. During the last 30 years, a wide network covering almost all states and union territories has been established. This growth in physical establishments–buildings, facilities, equipments, trained manpower etc. has put the Council on a solid base by the turn of the century.

However, from the beginning of the 9th five year plan period in 1997, main priorities were fixed on consolidation, quality and efficiency through optimization. The Council is currently operating from 29 locations and it was decided in 2000 not to bring new science centres developed by NCSM under the fold of the council for running and maintaining them. On the other hand, the present growth is being sustained by making the existing units of NCSM more modern, effective and technology-conscious and by establishing Science Centres in hitherto unrepresented areas on a turnkey basis with NCSM expertise and handing them over to the state or local authorities after inauguration. To maintain quality of activities conforming to the NCSM norms, continuous monitoring is done and full technical support is given by the Council to such centres handed over to local State authorities. One of the major initiatives taken in the 10th plan (starting 2002) period is establishment of Science Centres in all the States in the North-East and also in the Andaman and Nicobar Islands. These modern Science Centres in geographically remote areas have now become hubs of scientific activities for the local communities in general and the students in particular. This way, the science centre movement is becoming regenerative and the target is to give the movement decentralised local bases. Thus, the spread and penetration of non-formal science education is expected to receive a wider and stronger thrust in years to come.

3.0 Evolving Qualities

3.1 Evolving Exhibits

The first science museums in India followed European models and set up the galleries with artefacts, working exhibits, scaled-down models, cut-section models, dioramas etc. supported by charts and flow diagrams. Till early '70s, the entire NCSM team gained expertise in planning and development of high quality working models and cut-section models. Most of the exhibits were mechanically or electro-mechanically operated and some had simple electronic control using analog circuits. The advent of interactive exhibits in the US and their use in explaining scientific principles greatly influenced the exhibit designers of NCSM. During late '70s and early '80s designers from NCSM visited science centres in the US for studying interactives developed there. At the same time, the Indo-US Sub Commission for

Exhibits in Galleries

Exhibits in Galleries

Education and Culture, an inter-governmental body supported exchanges of staff between US science centres and NCSM science centres. Very soon the Indian designers started producing interactives on a large scale on various disciplines of basic sciences. Most of them were found to be immensely successful in teaching fundamental principles of science to the visitors. NCSM used this newly found expertise in developing all the new science centres founded since the '80s. The Nehru Science Centre in Mumbai was the first to be called a *Science Centre* and was quite different in character from the Science Museums operating in Kolkata and Bangalore by that time. Since the beginning of '90s, digital electronics, computer based exhibits and multimedia supplements were introduced largely to replace the analog controls, charts, translides, etc.

Travelling Exhibitions

To cope up with the modern trends of exhibit design and display, NCSM decided to set up a strong exhibit development and training unit. The purpose of this Central Research and Training Laboratory (CRTL), founded in the NCSM headquarters at Kolkata, was to develop new exhibits, controls, animation techniques and display techniques used in state-of-the-art exhibits. At the same time, training modules were designed to train staff at various levels working in the Council. The training involved all the technicians, educators, designers and also senior management staff. During the last decade, the capability of development of different kinds of exhibits–interactives, computer-aided, image processing based and immersive, has been acquired by the CRTL. This expertise in exhibit making received international acclaim and many science centres outside India started acquiring their exhibits from NCSM on a regular basis. This is also the time when NCSM started developing new science centres abroad on a turn-key basis. At present, NCSM is known in India as the nodal agency for science centre activities and the Council is not only developing new science centres with the government funding but also setting up centres for various corporate bodies in different places of India.

3.2 Evolving Human Resource

The Curators and exhibit designers working in science museums in the early days got exposure in various European science museums and received training in science museum activities. Later on, when interactives became sought after objects in the profession, such curators and exhibit designers were trained in this area.

Effectively, NCSM as a whole shifted its focus from the science museum arena to the science centre field.

Most of the new science centres became activity oriented and hands-on in nature and therefore, called for a new kind of training for the staff manning such centres. In India, the science centre curators are recruited from the engineering disciplines or from science disciplines and are exposed to design, development and fabrication of exhibits on the shop floor for the first few years before they are given specific assignments.

However, as for sustaining a high class network, expertise in exhibit development need to be supported by expertise in evaluation, activity design and management techniques, a need for a formal professional course for the curators and exhibit designers was felt. To meet this need, NCSM started a formal course leading to MS Degree in Science Communication from 2005 in collaboration with BITS–a leading engineering university of India with faculty support from the Smithsonian Institution, USA. The students of this two year 4-semester course are taught various modules for developing a broad perspective of the profession. Through the assignments and project work during the course, they also apply the knowledge gained in the classes to the real field. As a part of the course, exchange of staff between the NCSM and the Smithsonian Institution takes place regularly. Such exchanges have been funded by the Indo-US Science and Technology forum.

For senior level persons in managerial capacities, project management training is given to them through courses developed by renowned management institutions in India. As the demand for new science centres in India is growing, these management courses have been found to be very effective to handle new projects.

4.0 Impact Assessment

The main objectives of NCSM are to popularize science and to create a scientific temper among the public. It being so, the end-products are somewhat intangible and difficult to measure. However, following the internationally accepted procedures, continuous assessment of NCSM's specific activities and its impact on the communities being served, are being done. Along with such studies, visitors survey conducted in all the centres generate data for managing the centres. NCSM offers

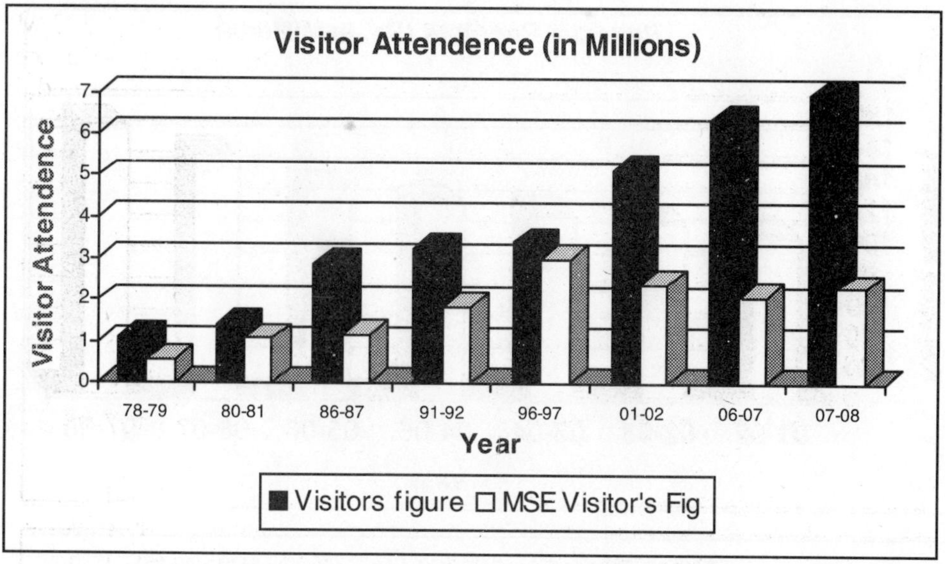

fellowships for conducting impact assessment and it is expected by end of 2008, a substantial database would be available to carry on the assessment work of NCSM further.

5.0 Getting Ready for the Future

The NCSM continuously strives to keep its science centres abreast with the state-of-the-art techniques. As technology changes, NCSM tries to acquire new equipment and adopt new techniques to produce modern exhibits which are efficient in science communication. For example, with the fast development of communication through internet in India, NCSM is preparing its centres to proliferate Internet-based science teaching. Specific training is being given to its staff and required equipment is being procured. It is expected that such Internet-based activities would be widespread as soon as the broadband facilities in India are made easily available at the receiving ends.

Mobile Science Exhibition Buses

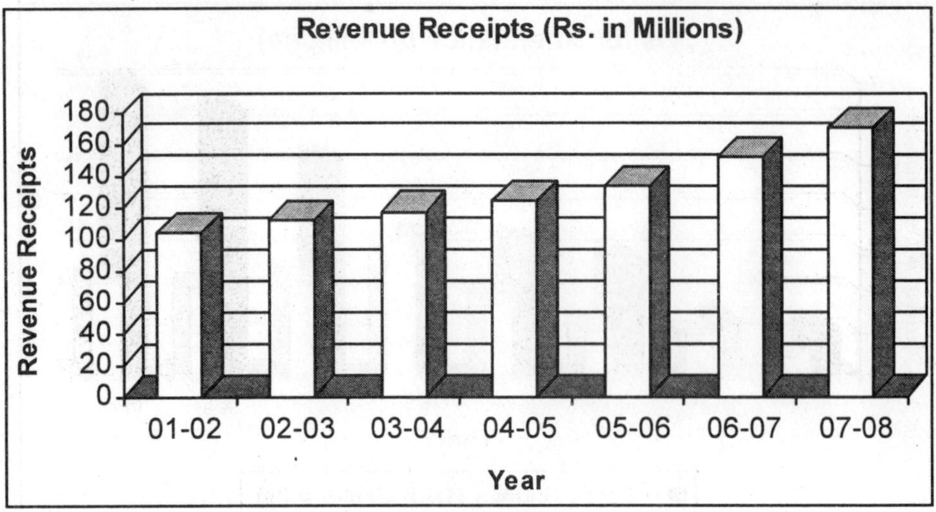

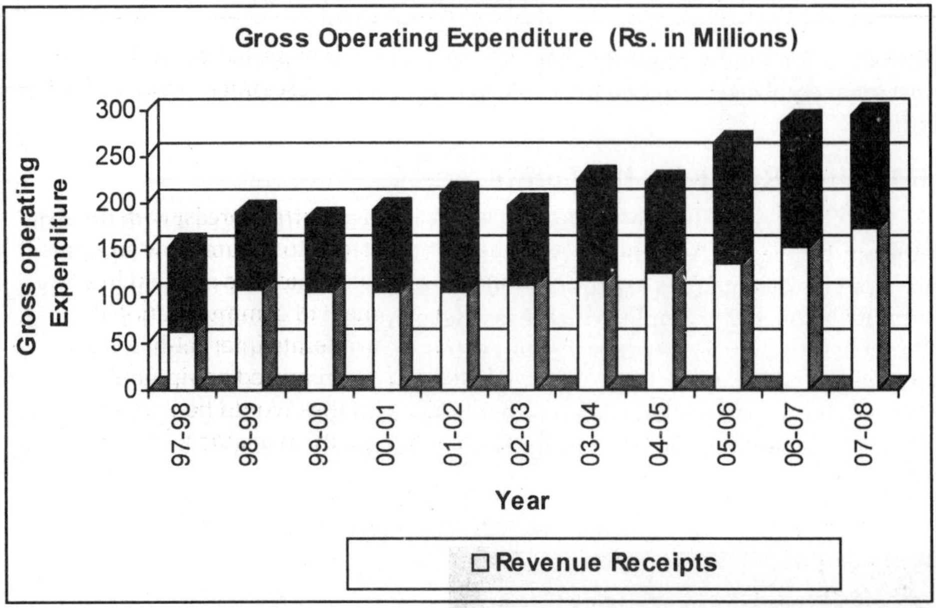

About 70 per cent of the Indian population live in villages. The Mobile Science Exhibition units described earlier try to reach them as much as possible but in a vast country like India, a huge fleet is required to serve the purpose fully. NCSM at present runs 22 such Mobile Science Exhibition units called Museobuses and for 9 months in a year they tour the rural areas taking the message of science to the rural folks. Recently, some of such MSE units are being fitted with antenna to receive programmes being uploaded from major science centres of NCSM through the satellites.

India has a large area covered with high mountains, deserts or forests. It also has a number of islands under it. People living in such remote places need modern science

Science Fair

Science Seminar for School Students

Science Drama

Taramandal–Portable Planetarium

education. During the last five years, NCSM took an initiative to set up new science centres in all these hitherto unrepresented areas and brought them under the science centre network in India.

The science centres operating under the NCSM generate more than 50 per cent of its total operating budget through its revenue earning. The balance is provided by the government. The Science City in Kolkata, however, runs on its own steam and it has remained financially self-sustained since the day of its inception in 1997. This is considered as a major achievement in the non-formal science education field.

6.0 Redefining Objectives

The main objectives of NCSM which were set 30 years ago need a re-look now. As a matter of fact, many of them are difficult to be quantified as they are somewhat intangible. The intention of the Council should, therefore, be redefined and the deliverables are to be made reasonably quantifiable.

India is passing through a transition phase when the resource based economy is being gradually converted into an industrial economy or a knowledge economy. However, to sustain this, it is felt that the element of innovation must be impregnated in all types of education–particularly science education. The role of science centres in developing an innovation bias in the minds of our youth and in the education system as a whole will be important. Science Centre activities must be related to the movement for innovations in all disciplines of life.

7.0 Science and Technology Policies of the Government of India

Public awareness of Science and Technology constitutes an important part of the latest S&T Policy (2003) of the Government of India. While formulating the relevant strategy for achieving public awareness, the Policy notes that

- There is growing need to enhance public awareness of the importance of science and technology in everyday life, and the directions where science and technology is taking us. People must be able to consider the implications of emerging science and technology options in areas which impinge directly upon their lives, including the ethical and moral, legal, social and economic aspects. In recent years, advances in biotechnology and information technology have dramatically increased public interest in technology options in wide ranging areas. Scientific work and policies arising from these have to be highly transparent and widely understood;
- Support for wide dissemination of scientific knowledge, through the support of science museums, planetaria, botanical gardens and the like, will be enhanced;
- Every effort will be made to convey to the young the excitement in scientific and technological advances and to instill scientific temper in the population at large;
- Special support will be provided for programmes that seek to popularize and promote science and technology in all parts of the country. Programmes will also be developed to promote learning and dissemination of science through the various national languages, to enable effective science communication at all levels.

In general the activities of science museums and science centres have been identified as an area for promoting public awareness of science and technology and the Government of India has given a special thrust in this area.

8.0 The Challenges

8.1 The Decline of a Career in Science

Till the '80s, the best of the high school students opted for scientific streams for higher studies. With the economic liberalization in India and the influence of globalization, an overwhelming majority of students with high quality performance now opt out of science streams and choose engineering and other technical areas for higher education. Naturally, there has been a serious decline of career in science particularly among the talented young people. During the period 1950-2000, enrolment in the undergraduate level has increased more than ten fold. Despite this growth, the overall percentage of high school students opting for science in the university undergraduate level has consistently declined–from about 32 per cent in 1950 to less than 15 per cent in 2000. It is true that in India at present a career in engineering or information technology is highly rewarding and a career in science is by far less attractive in terms of remuneration, career growth and other prospects. Although this is the major reason for a decline in career in science in India at present, there are other reasons related to the formal undergraduate science education programme which lacks excitement of learning in science. Science Centre activities can very effectively and substantially fill up what is lacking in the formal science education system and possibly bring back the thrill in science learning through an interactive *doing science* approach.

9.0 New Initiatives

Since the beginning of the 21st Century, the NCSM has been functioning with 29 operating science centres and science museums located in different parts of India. The activities of the NCSM network is so popular in India that demands for setting up new science centres have been coming regularly to the NCSM and the Department of Culture, Government of India from many State Governments. The growth of the infrastructure and its subsequent maintenance under the administration of the NCSM were very carefully examined to determine the nature of future growth of the science centre movement in the country. It was felt that administering a network larger than the present one may prove difficult for NCSM. But, at the same time, the messages of Science and Technology need to be communicated to areas hitherto unrepresented by NCSM in this vast country with population more than 1 billion. It was decided that NCSM would set up science centres in different parts of the country particularly in the States and territories which so far do not have any science centre and later on, hand them over to the respective local authorities or societies formed by the local authorities for the purpose to operate and maintain such new science centres.

On the basis of this decision, work for setting up new science centres started from 2001 in the North-Eastern parts of the country, island territories and also in the States which do not have science centre so far. All these science centres have been completed in time and are now in operation. As a part of the package offered by the

NCSM to State Governments for setting up new science centres, employees engaged by the State Governments to operate the new science centres would be fully trained by the NCSM and technical support from NCSM would be made available to the local authorities of State Governments even after such science centres are inaugurated and handed over to them.

The growth of the NCSM network has been fast during the last three decades. Quality and consolidation of the network are the key challenges now. It is also important now to make the movement regenerative at the local level as the demand for new science centres and planetaria from different parts of India are continuously increasing. So far, the NCSM network has grown solely on government fund. But the mission of taking the message of science to nook and corner of the country would be possible if corporate sponsorship and private endowment funds are made available for this purpose. Fortunately large corporate houses and public sector units have started expressing their interests in setting up science centres. During the last three years, such contribution to NCSM from corporate bodies and other sources has crossed INR 75 million.

Chapter 2

The Role of Science and Technology in Developing Countries

Joaquin Fargas
Art Science Technology, Argentina
E-mail: joaquinfargas@exploratorio.com

ABSTRACT

Science and Technology has been along the history of the human civilization a way for the development and had improved the quality of life. As side effect sometimes they wide the gap between the people around the globe and many times were used for bad. The purpose of this presentation is to show briefly how S&T could help the developing countries with emerging economies, bridging the gap between th_ poor and rich communities and how S&T could be part of the solution for global issues like: global warming, famine or overpopulation.

Keywords: *Science centres, Emergent economies, Developing countries, Knowledge society, Art and science, Environmental education.*

1.0 Introduction

Science and Technology and the Human Civilization

Technology and science started at the very beginning of the human civilization. As soon as the man started questioning himself, science emerged, and as soon the man realized that something in the environment could be modified by their own action, technology started, and also realized that he could enhance the performance of the technology by studying how it works, making it more predictable, making science and technology inseparable. In many centuries, science and technology grew together, influencing each other.

Could any of us imagine a world without Science and technology? Could we imagine any progress? What about the quality of life? What about the future? Is there any hope for the next generations without science and technology? It is impossible to imagine a future beyond science and technology. The human civilization is facing many global issues that must be addressed in many ways but always including science and technology in the solution. *"Problems cannot be solved by the same level of thinking that created them"*–Albert Einstein. It means that we need to use our improved knowledge to find a solution; we cannot stay still just watching as time goes by, in this case our destiny as human civilization is extinction.

Global Warming and Climate Change

For the experts climate change is already happening. Global Warming is not just affecting the poles, causing a big thaw, but making the climate to change. Now it is more than an environmental issue, it is an economical issue. The insurance companies say that the annual cost for them, because of the climatic disasters, is more than 40000 Million US dollars since the last three years.

We do not know exactly, what is the role of S&T, if we can find a solution through them, but for sure it will help. It would help finding other sources of clean energy or using in a more efficient way the existing ones. Hydrogen, Fuel cells, Fusion Reactor could be the real solution for energy in a short time, but only if we invest in Research and Development (R&D). The constant demand of food cannot be solved if we do not apply biotechnology to crops and cattle, Science and Technology is everywhere.

2.0 The Role of Technology in Developing Countries

Several developing countries rely their future in their commodities such as crops and minerals. It will take many years to reach the level of the developed countries if they pursue doing this. The fast track is technology and science.

Soya Crops

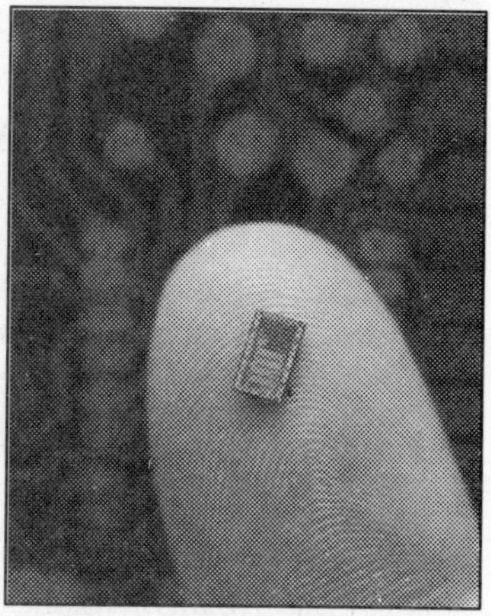

Chips

The picture shows soya in Argentina. The crops will reach a surface of 50 Million acres.in 2009, (source MDZ Online). Many countries, rich in natural resources, invest few money in Science and Technology.

This unknown chip on the figure could be from a factory in Singapore. This small country, a real city-state, is investing in education on science and technology. The best way to continue growing.

The cost of a Soya ton is US$250 the same as the last chips for computers

1 Ton (Platinum Raw Material) 350 M Rands

1 Ton (Platinum Manufactured into Jewelry) 850 M Rands

The comparison of the cost of raw platinum and manufactured with just simple technology makes the difference.

3.0 Towards a Knowledge Society

The development of a Knowledge Society is very important for increasing the growth of a nation. Everywhere people with background on Information and Communication Technology (ICT) is required, knowledge leads to better jobs and social inclusion.

Where Does the Knowledge Lives?

Knowledge and expertise reside not at the Universities, not at the libraries, but in the people's mind, therefore knowledge moves, every time a researcher or scientist transfers his knowledge. It is very important to encourage these people to stay in their countries, but they need opportunities to work, not just a fair wage but the logistics and support to do their research, to publish, to enhance their academic level. Opportunities must be given for the native scientists and researchers.

4.0 The Role of Science Centres

Science Centres play an important role on society, teaching vital issues for the quality of life, rising awareness on the importance of science and technology for the growth and encouraging youngsters to follow careers related to ICT.

Hands on experiences in science could change the life of many young people. It is necessary to be involved in real science to make decisions about what careers are good for someone to follow. Science, engineering and technology are not-among the easiest careers to study. Learning medicine, biology, nuclear engineer are in some way difficult and the students must be interested or motivated to follow them.

5.0 Our Contribution

5.1 The Exploratorio Art, Science and Technology Centre

The Exploratorio, Art, Science and technology centre in Argentina has been working for the popularization of science and technology since 1990. Its mission is to improve the quality of life and help in the preservation of the planet Earth for the future generations, through raising awareness on science and technology. As an example we will explain the outreach program "Facing the Future (www.exploratorio.com/mirandoalfuturo) related to alternative energies, the Biosphere (Project and Sunflower, Sentinel for Climate Change related to environmental issues.

5.2 Facing the future

Description

Following the statement "The earth has not been inherited from our parents but borrowed from our sons", we develop this program based upon the concept of "saving" not only related to money but in a more extended area. Taking good care of nature and its energy resources, acknowledging the concepts of "saving" which guarantee a better standard of living without jeopardizing the future of the planet or even our own. The Program emphasizes on the importance of the sun as an everlasting source of energy that will help to stop pollution, preserve the woods and fossil resources and, in that way, lead us to a better quality of life.

Key Issues

- Climate Change
- Deforestation
- Desertification
- Rural Poverty
- Environmental Education

How it Works

The program uses a combination of improved solar energy technology, materials technology and educational technology. Electricity to schools is provided through installation of Solar Panels. Materials are brought to the schools and the communities to construct a solar oven. Technology and education both together lead to a sustainable development.

5.3 The Biosphere Project

Description

Spheres, hermetically shut, contain the necessary ingredients, the basic elements for life to happen. Inside, they hold a complex ecosystem, which needs solar energy to

Facing the Future in Bacoya, Salta–Argentina

Solar Panels

Solar Oven

subsist and develop. Biospheres are the representation of live forms in the planet. They reproduce the basic process of functional life. In this way, the spheres act like the atmosphere in our planet, they are the limit that joins the outside and the inside. It configures a closed system, like Earth. The only thing that receives from the exterior is the solar energy: the beginning of the life chain. The plants of the ecosystem grow by transforming light and nutrients into their own food, via the process of photosynthesis.

In these pieces, the significance falls in the notion of exchange, an interaction between the art craft and the surrounding. The vitality of the biospheres will depend on the situation in which the pieces are implanted. Variables like temperature and the amount of sun light would determine their survival.

Key Issues

- Climate Change
- Ecosystem
- Environmental Education
- Care Art

The World in our Hands

Micro Biosphere

Life inside

(http://www.joaquinfargas.com.ar/blog/?cat=6)

5.4 Sunflower, Sentinel for Climate Change

Description

Sunflower, Sentinel for Climate Change puts forward scientific concepts and theories in a poetic and ludic way, and urges people to understand the properties of nature and the need to care for and preserve them.

The piece, an "intelligent" metallic sculpture that emulates the behaviour of a sunflower, opens its petals at dawn and follows the sun during the day. The petals

Sunflower Installation–Ushuaia, Tierra del Fuego–Argentina 2007
www.sunflowerweb.com.ar

are solar panels that generate the necessary electricity for the functioning of the flower and its lighting at night. Moreover, through video cameras, the flower captures images of the surrounding landscape, the sun, and itself on its daily rotation and tilting movements.

But besides its aesthetic function, the piece operates as a weather station, monitoring the environmental conditions, such as air pollution, UV radiation, different kinds of temperatures, and so on. Then, the information collected by the flower is transmitted through wireless systems: the data is sent both to a module placed inside the institution acting as the project–"host" and through the Internet, where the images taken by the flower as well as the weather data are kept online.

Although the first Sunflower was installed in the city of Ushuaia, province of Tierra del Fuego, Argentina (the southernmost city in the world), this project has a global nature, not only by providing access to the information collected by the flower anywhere in the world, but also because it contemplates the installation of other flowers in other cities worldwide.

Mission

Sunflower proposes a kind of communication aimed towards a new conception of education that views contemporary art–and the channels through which it circulates–as a *space* and a *time* in which to develop constructive criticism about our context.

Key Issues

- Climate Change
- Environmental Education
- Global communication

6.0 Conclusion

There is no doubt about the important contribution of the Museums and Science Centres for the formal education, the quality of life and the growth. Every country must have a sustainable program related to this field.

Bibliography

Association Ciudadania Solidaria, Buenos Aires, Argentina.

Bioart Lab, Maimonides University, Buenos Aires, Argentina.

Centro Científico Tecnológico Exploratorio, Buenos Aires, Argentina.

Foundation Espacios Verdes, San Luis, Argentina.

Foundation CIENTEC, San Jose, Costa Rica.

Red Pop, Network for the popularization of Science and Technology for Latin America and the Caribbean, UNESCO.

Part II

Evolution of Science Centres

Chapter 3

Social Appropriation of Science and Technology

Elizabeth Hoyos

Maloka Corporation, Colombia

E-mail: ehoyos@maloka.org, elizabethoyos@gmail.com

ABSTRACT

The purpose of this paper is to analyze some aspects of the economical and social impact of science centres as catalysts of the social change that is urgently needed. We will refer to the leadership of Maloka in the Colombian experience. The growing influence of science and technology (S&T) is significant to our welfare and it is the driving force in the development of mankind. The most important challenge for developing countries is to prioritize the generation and application of knowledge and innovation in relation to their social challenges and opportunities. Colombia, a country that has been through serious economic and social crises is in need of commitment, policies and actions from our leaders, that will ensure successful articulation of globalization in order to provide better opportunities for the human development of individuals and communities. The ability to make decisions, both, at individual and collective levels would open opportunities in accordance to our talent, resources, needs and opportunities which would improve the quality of our lives. Such a process should begin with government and private leaders who should recognize the role of science and technology as a catalyst for cultural and economic transformation. This kind of work should be a partnership aimed at making radical changes in formal education, starting at a very early age. The Colombian Association for the Advancement of Science created Maloka Science centre in December 1998. It is a private non profit self-sustaining organization, whose purpose is to contribute to the social Appropriation of Science and Technology–AST-, concept that in Spanish means to make something your own, transforming attitudes and actions to integrate these topics as part of daily life and productive processes. As an institution, Maloka has unfolded different strategies within the frame of non-formal education. These strategies look toward the development of experiences which encourage learning processes in communities with differing characteristics, significantly impacting their relationship with knowledge and simultaneously stimulating the development of new programs in other regions of the country. Maloka sees itself as an agent of social articulation around science and technology which, beyond attracting a considerable number of people to its different scenarios, is focused on building meeting inspirational spaces, and forming networks to extend the impact of the institution.

Keywords: *Maloka science centre, Science and technology, Institution, Development, Knowledge.*

Introduction

Knowledge enables us to make informed, democratic decisions and to exercise the power of self-determination. It is paradoxical that our knowledge–based society faces the challenge of consolidating a community where scientific capabilities, technological advances, and innovation contribute to this freedom and to socioeconomic development, while some political and economic policies prevent this freedom from being exercised due to criteria, regarding inclusion and equality on a world-wide scale.

Science and Society

In order to study the possible alternatives for harmonious development in our society, it is very important to analyze and articulate the internal historical process of Science since the XVIII century, when this topic was in the hands of a small number of individuals (experts, as well as amateurs), who were carrying out their scientific activities privately, without receiving anything in exchange. Economic development was the net result of intellectual interests on the part of individuals who dedicated time, inventiveness, and money to understanding the phenomena in their surroundings.

In the XIX century, political, social, and technological change made governments aware of the economic and geo-strategic consequences of development in the fields of science and technology (S&T), pushing them to adjust educational and institutional models activities in these areas. Systems were created to professionalize leaders in the field of science in order to satisfy the needs of a small community from the academic point of view. Disclosure of information was carried out as a social activity addressed to select groups.

In the XXth century, interests with regard to these topics broadened, as private companies turned into multinational concerns, structured on the trade of products based on Science and its technological applications. Profits from scientific and technological activities began to become a priority, and the differences between developed and underdeveloped nations deepened and became more evident. Parallel to the advancement of knowledge, the concept of spreading and popularizing science and technology was conceived as one-way message from the experts to a general public, considered to be a group of passive and ignorant receivers.

During the second half of the 20th century, a clear concept of a society based on knowledge is gaining acceptance every day and opening up to the interests of private companies. Multinationals are consolidated and the differences deepen between countries that do not invest in research and technological development and those that do so, thus appreciably raising their socioeconomic levels thanks to the products and services generated by knowledge and its applications. This difference emphasized the need to address topics of common interest, aimed at solving problems and developing the potential of less privileged countries, awarding scientific and technical development the pertinence and coherence required and adapting it to local needs.

Sharing this strategic information with the community gave way to the need for the creation of awareness and critical and proactive attitudes among citizens regarding these subjects. Opportunities arose both for continuing to concentrate power, as well as for facing the major problems of humanity in fields such as health and agriculture. However, quality of life indicators began to show the urgency to transform developing countries regarding these topics.

Paradoxically, while developed nations are the greatest generators of knowledge, they have also become the greatest predators of the environment. This situation requires the rethinking of the conditions in which science and technology are generated and appropriated, since it seems to be clear that only from systemically structured strategies and policies that are socially and environmentally responsible can environmental harmony ensue, with knowledge as its basis. This implies creating a common front among scientists, political and entrepreneurial leaders so that the decisions that are made will be appropriate for the quality of life of our nations.

Hence, the idea of a society based on knowledge must move toward the constitution of a community where scientific capabilities, technological development and innovation contribute to social and economic development under the criteria of inclusion, equality and environmental responsibility. This implies the involvement of the entire community in a revolutionary concept of active scientific culture, whose characteristics must allow democratic access to it with the possibility of free generation, utilization and distribution of knowledge. An example of this would be the cooperative scientific model (Barrio, Alonso, 2008) which is characterized by actively involving the citizens in processes of generating socially useful knowledge which also contributes to the individual and collective satisfaction of creative development. In this model, knowledge and action join in efficient contribution, breaking the barriers between professionals and amateurs.

The Path to Science Directed by Ethical and Democratic Values

Humanity should reflect on the need to build and reinforce ethic and moral values in order to achieve a knowledge-based society. Science and technology can be misused exclusively to the advantage of a particular elite and of special economic interests, with disregard to the deterioration of the community and of the planet as a whole. Today there are many questions about bio-fuels, stem cell transplants or genetically modified foods.

Science, technology and innovation are also very important in the generation of sustainable development strategies, and for actions aimed at urgent and complex challenges such as Global Warming. It is imperative for every citizen to become an active part, not only in the solutions of domestic or regional problems, but also in the conservation of our planet.

The role of citizens should be redefined as active participants, implicated within an integral system, where science and technology appear as transversal topics. This presupposes the design of innovative strategies aimed at the systemic transformation of culture.

While respecting the concepts of disclosure or popularization used all over the world, it is important to go beyond a one-way relationship between science and society. With this in mind, in Latin America and specially in Colombia, we decided to look for a wider concept, a Spanish term that makes a lot of sense. We chose the word APPROPRIATION which means to make something your own. In terms of science and technology, it implies the creation of awareness, a critical and proactive spirit, the transformation of attitudes and actions with the vision of making these subjects part of the daily life for the welfare and collective prosperity of all citizens.

Our goal is the construction of a society that generates and appropriates knowledge, and whose members can use their learning capacity throughout their life; an innovative, visionary and free community that can develop its potential, talent and natural resources to achieve sustainable development.

We wish to conclude with the paradigm stating that the main actors in terms of knowledge–scientists and innovators - have a one-way relationship with a passive community who merely acquires information without participating. It is imperative to build public opinion, so citizens can exercise their rights and duties in the decision-making processes affecting their future. We propose a society in which the learning process is continuous and universal–Life-long Learning-, where adults also take advantage of their learning capacity.

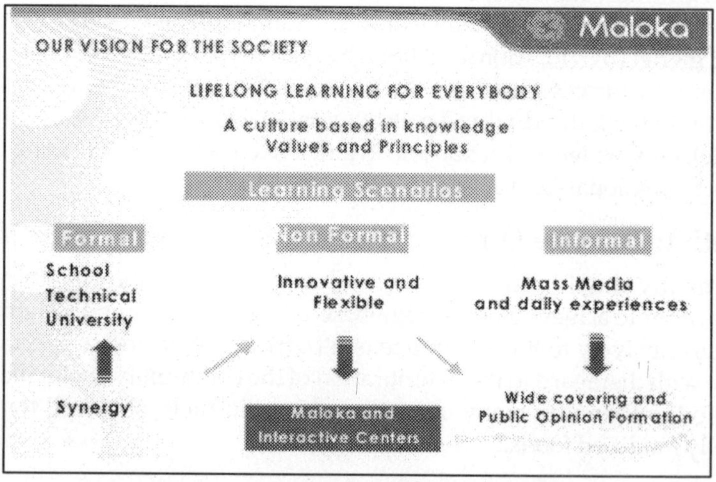

Figure 3.1: Life Long Learning Process

Social Appropriation of Science and Technology (AST) is a concept applied within a *non-formal education* environment. It is transversal because it reinforces and complements formal learning (school, technical and university), while having a flexible structure that can be evaluated. It articulates and mediates informal learning mediated by the mass media and daily experiences. Interactive, significant personal experiences are the key pedagogical tools for the AST.

AST has an important role in sustainable development as well, allowing the generation of efficient economic and environmental conditions. It also contributes to innovative processes for the production of goods and services, in order to improve competitiveness in a globalized economy.

Consequently, in AST there are several actors: general society, leaders and decision makers, civil organizations, the productive sector, journalists and the media, formal and non-formal teachers, etc.

Colombia and the Appropriation of Science and Technology (AST)

Context

Colombia is a democratic, pluralistic and participative nation, located in the northwest tip of South America with an area of 2.070.408 km^2. It is the only South American country that has coasts on both: the Pacific and the Atlantic oceans (through the Caribbean Sea). It has a population of 44 million, 64 per cent of which is racially mixed and the official language is Spanish. It ranks second highest in the world for biodiversity due to geographic and climatic conditions.

Figure 3.2: Latin America and Colombia

Colombia has little understanding of the importance of science, an issue that directly affects the quality of life and the decision-making capacity. The investment is only the 0.4 per cent of the GNP in research and development. Its technological development is obsolete in the productive sectors.

Education on S&T is deficient and generates a meagre interest in scientific and technological careers among youth. There are also a low number of full time researchers and engineers dedicated to technological development. Additionally, adaptation to the evolution of S&T has been jeopardized due to lack of proper resources on the part of educational institutions. Nevertheless, there has been an impetus toward the transformation of the system in the last fifteen years.

The term Social Appropriation of S&T–AST- was first used in Colombia in the Mission for Science, Education and Development summoned by the President of the Republic in 1994.

Several proposals were made and the most relevant were those related to AST. The concept of Interactive Teaching and Learning reached Colombia twenty three years ago by way of the Science and Games Museum. This exposure allowed the creation of a national network called The Lilliput Network.

Maloka, a private non-profit organization, is a national program with international projection that has received more than 18 million of visitors including Colombians and foreigners, becoming as one of the icons of development not only in Bogotá but also in the rest of the country thanks to its flexible and universal character: Culture, education, science, technology, innovation, environment, recreation and tourism that contribute to the construction of a "society and economy of learning, knowledge and innovation".

Through different strategies and a non-stop commitment to fulfill the philanthropic mission of integrating the science and technology into the citizen's daily lives, has been able to achieve, during the last 10 years, a position as one of the most innovative, transparent and efficient institutions in Colombia.

Years later with the birth of Maloka, a national movement was started thus promoting the creation of Interactive Centres around the country with its most recent addition, the Interactive Centre of Medellín, Explora. This Institution has joined the family of Interactive Centres which have become agents of change in Colombia. We, scientists of the Colombian Association for the Advancement of the Science, worked in strategies for AST, that finally were focused on science centres.

Interaction Science Centres as AST Strategies

We believe that their frameworks are not only settings where knowledge and learning are continuously experienced but also ideal interactive scenarios and therefore agents of social change. This statement is backed by Maloka corporation as an organization dedicated to the social appropriation of knowledge for the last 10

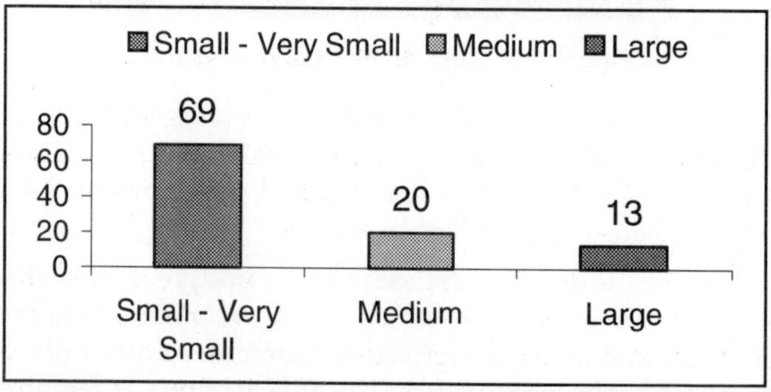

Figure 3.3: Distribution of Museums and Science Centres in Latin America According to their Size (2003)

years. From this experience we propose our viewpoint regarding the social impact of interactive centres, clarifying that we are considering an experience within a specific context and cannot, therefore, make a generalization. What we can suggest as a general principle in terms of the level of impact that these scenarios can generate, is their direct impact the context, the dialogue with the diverse audiences, and their capacity to interpret and thus begin to transform their own reality.

In Colombia as well as in Latin America there have been different initiatives aimed at establishing interactive centres. This tendency has been increasing during the last few years, as can be seen in the Figures 3.3.

Maloka Corporation: A New Dimension for Colombia

Maloka: A National Symbol

Maloka Corporation was created in December of 1998 as an initiative of the Colombian Association for the Advancement of Science with the support of the District and National Government and the private sector as well, thanks to a policy that allows public and private institutions to join efforts in the development of common projects. It is a self-sustained, non-profit private organization.

Maloka, an indigenous word, means the centre or the central house of our native tribes, where they receive wisdom from the Universe; a place where the most important decisions are made and where the community finds rest and relaxation. It is used in 5 countries in the region and illustrates our origin and the importance of our cultural cohesion. Our Maloka of the XXI Century, *THE FASCINATING ADVENTURE OF KNOWLEDGE*, integrates encounters that promote the growth and enrichment of the community, where the human race enjoys its due respect. It is designed to articulate social and human sciences in a local coherent context. Its logo is a spiral galaxy with the human being in constant evolution and in harmony with the environment. We develop different experiences geared toward the meaningful learning of science and technology aimed at diverse audiences.

Our Mission is to contribute to the Social Appropriation of Science, Technology and Innovation by generating a passion for learning, critical spirit, and a proactive

Figure 3.4: Indian Maloca

Figure 3.5: Maloka Science Centre

sense as well as an awareness of the transcendence of these factors in the social, economic and cultural development of daily life for all Colombians.

Our Vision is that by 2011, the Maloka Corporation will contribute significantly to the enrichment of Colombian culture through the generation and social appropriation of knowledge and innovation; it will be an outstanding organization, based on principles and values, well respected in national and international scenarios, with a solid financial structure and highly qualified and passionate human resources.

Maloka has become a symbol of the community and one of the best known institutions in terms of S&T, above universities and research institutions according to the most important study in our country about the perception of Colombian people on these subject. (Colombian Institute for the advancement of S&T, Colciencias 2005).

Maloka was born as a national program with an international projection, and its most visible scenario is the Interactive Centre that carries out actions within its three worlds:

The *Maloka Interactive Centre* is a transparent underground 17 000 m^2 building. At present we are initiating the construction of 23 thousand additional square meters. The Centre has 10 exhibit halls related to diverse subjects: the City, Water, Petroleum, the Universe, Biodiversity, Life, Telecommunications, Human Beings, S&t in Colombia,

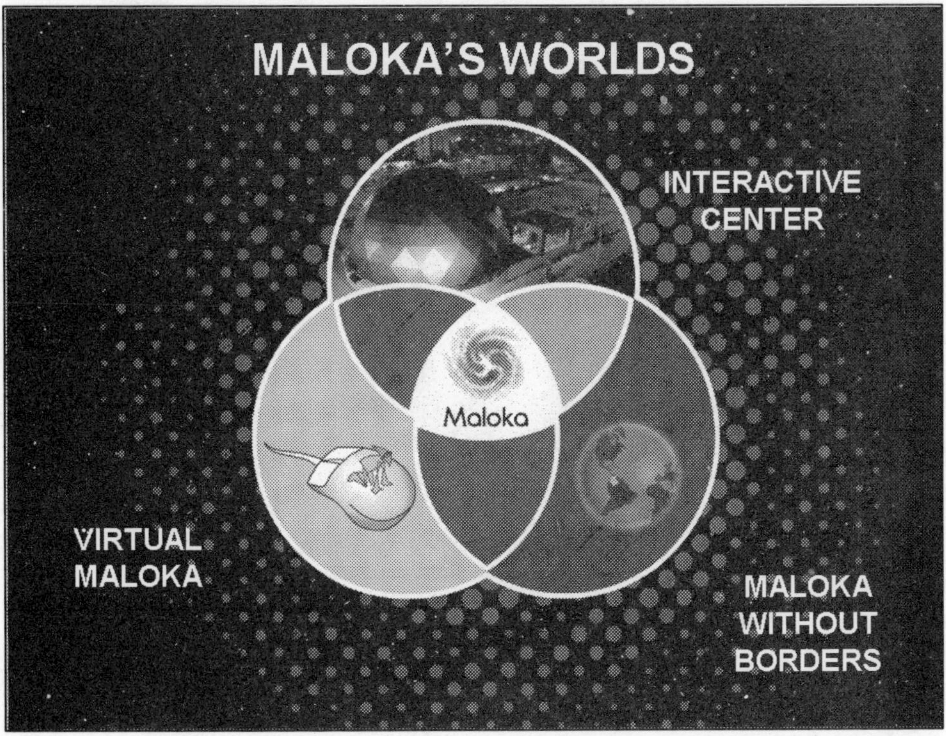

Figure 3.6: Three Maloka´s Worlds

and a special area for small children. The exhibit halls have more than 350 interactive modules designed and built in Colombia. We have the first large format theatre in South America; its programs complement the other experiences offered by the Centre. We feature scenarios for educational activities of diverse characteristics. In our mechanic and electronic workshops we design, produce, and maintain the exhibits. We organize forums, debates, courses, and the S&T Clubs for children and youth for ages of 4 to 14. These groups meet weekly at the Centre.

Virtual Maloka–www.maloka.org–was created in 1997 as a setting for children, teenagers, adults and teachers. Its virtual and application modules are designed for cyber explorers to achieve fascinating adventures in knowledge. It hosts more than 13,000 visitors daily and has had numerous national and international acknowledgements.

Maloka without boundaries which includes Travelling Maloka, consists of 2,000 m^2 of travelling exhibits, designed with the main objective of reaching those remote regions in the country that lack the opportunity to have access to these experiences. We create alliances with regional leaders and train college students and teachers to assure continuity in related activities following the visits. We have travelled to more than 45 cities in Colombia as well as some neighbouring countries. We also offer Maloka and the Stars, to learn about astronomy in a fun and pedagogical way, and

Figure 3.7: First Large Format Theatre in Latin America

Maloka Goes to School, which reaches institutions with the purpose of continuing the process initiated in the Interactive Centre, within our permanent program of teacher training.

Strategic Alliances

Since its foundation, MALOKA has been a community-oriented organization that lives and grows according to the dynamic links among the different sectors of the community. Therefore it has created a network of strategic alliances from public and private institutions, such as the affiliates of the Maloka Corporation. The network envisages different options: voluntary advisors in Science Technology, Pedagogy, Communications, Management, etc.; sponsors and allies in order to develop strategic projects for the community.

Interactive Centres: Social and Economic Impact

Consequently, it has been our experience that Interactive Centres constitute the main tool for the social appropriation of Science, Technology and Innovation, since they are instrumental settings for articulating the multiple scenarios and contexts of society. They are much more than an enclosed space; they should be conceived not as a location but as organizations that generate significant impact at national and international levels with networks of programs that link their work to local reality, mass media, formal education and so on, while simultaneously providing ample coverage based on Maloka experience. Their impact is manifested in the following:

The formation of scientific thought and attitudes, structured by analytic logic to enable rigorous interpretation of reality as well as understanding of the complexity of relationships and situations that lead to appropriate decisions with consideration to social interests.

With the aim of structuring a clear proposal in these terms Maloka has developed action guidelines, depending on our targets, groups and scenarios for the construction of its proposals:

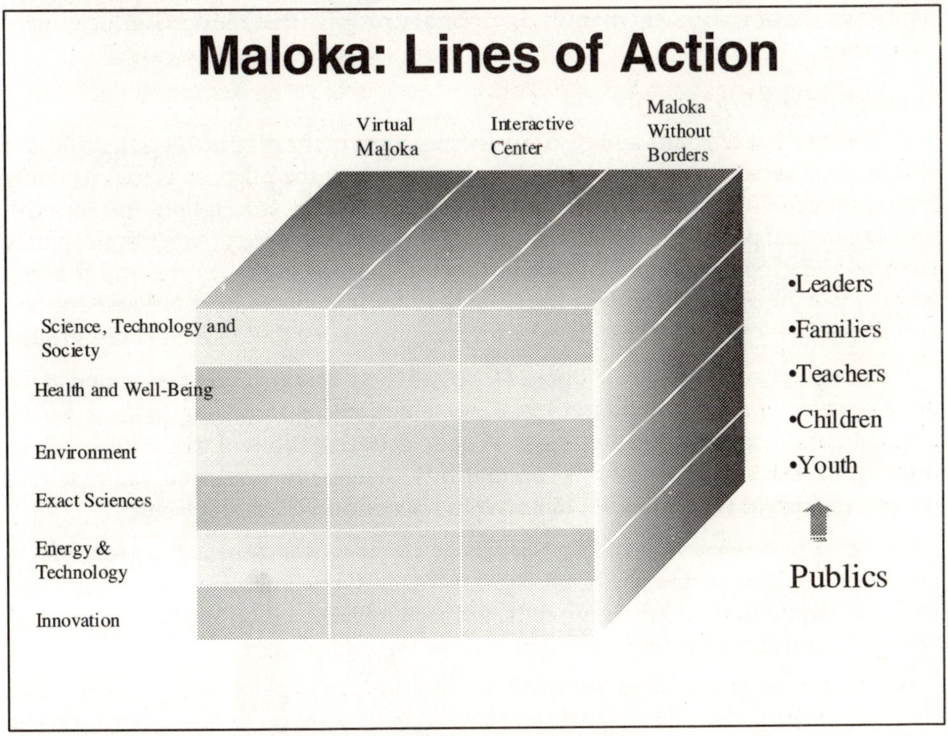

Figure 3.8: Maloka´s Lines of Action

Health and Welfare

This guideline is oriented towards the application of scientific and technological development for the prevention and solution of basic health issues to improve the quality of life. We also work on topics such as drug addiction and nutrition, the prevention of disease as well as social behaviour and ethic values.

Environment

This guideline is based on the use of knowledge of S&T as a tool for the improvement of environmental conditions. Work includes educational programs and project development with different communities in their territory. It is based on the understanding of a dynamic relationship among human beings, natural resources, and culture, and the analysis of local and global contexts as well.

Basic and Natural Sciences

We aim at shattering paradigms regarding science as boring and incomprehensible. We seek to arouse passion and a critical but proactive attitude for science through interactive strategies that stimulate curiosity, surprise and fascination in terms of understanding the world around us.

Technology

Emphasis is given to developing creativity through experimental and constructive processes; the central topics are robotics, air and flight, and their relation to daily life as well as the development of multidisciplinary projects that reinforce education in the country.

Innovation and Productivity

The creation of bonds among different members of the productive, scientific and educational sectors, with emphasis on entrepreneurs; the purpose is to effectively include technological knowledge in production practices throughout the country, and to generate permanent innovation processes to improve our competitive capacity in local and global markets. We are foreseeing the jobs of the future, and moving toward planning and providing the necessary tools to create demand and supply, and to meet the challenges associated with learning among the business communities.

Pedagogic strategies are proposed through these action guidelines, aimed at the development of scientific thought in diverse audiences according to their needs, interests, and cognitive development. We use different types of mediation such as interactive exhibits, workshops, animations, science theatre, and far reaching processes such as working with children in science and technology clubs.

1. *The development of new models of teaching and learning*, agreeable to contemporary cultural dynamics and the speed of scientific and technological advancements, without loosing sight of the social, cultural and economic impact of actions that are undertaken.

   Interactive centres are excellent locations to explore, apply, investigate, appropriate, disseminate and improve the discoveries and innovations on new ways of learning, with their respective strategies and different tools aimed at diverse audiences.

   Interactive centres facilitate acknowledging, valuing and working together with teachers in the construction of new teaching strategies that promote the recovery of self-esteem and a much deserved prominent position in society as well as supporting them in their processes in the classroom. The centres become a "home away from home" for teachers, a place of trust and support, where they can develop, evaluate, and implement their ideas.

   Interactive centres promote encounters of teachers with the scientific community and with political agents for the development of innovative strategies for schools as well as for other social scenarios with educational impact, such as the media and Internet, among others.

2. *The formation of critical opinion among all levels of leadership* for the purpose of strengthening the ties among science, technology and society, lending relevance to the processes in the generation and use of knowledge, beginning with the development of policies and strategies to consolidate the regional systems dealing with science, technology, and innovation.

In the fulfillment of their role as articulators and promoters of the concept of transversal knowledge, interactive centres are extraordinary scenarios for inviting actors with antagonistic but enriching points of view on significant issues of Science and Technology such as bio-fuels, stem cells or new forms of energy; everyone should not only learn, but understand and interact with the different points of view so that their own committed and proactive opinions can emerge as part of the transformation of society. Interactive Centres should promote ACCOUNTABILITY in terms of the rights of citizens and stimulate the development of the individual responsibility of *the citizen* thus breaking the paradigm that only political and private leaders are responsible for the decisions that affect the entire community.

The fact that multiple actors such as political and scientific leaders, the academic community, housewives, teenagers, and teachers are involved with the programs of the Interactive Centres brings about a NON SECTOR intellectual convergence, causing the results to have a significant and forceful impact on the communities.

3. *Inspiration for vocations in science and technology* with the intent to strengthen one's own capabilities in order to generate and apply knowledge.

   Interactive Centres are excellent venues for the construction of life projects for young people who are making job decisions as well as for the reinvention of possibilities for the workforce, based on the perspective of new occupations and professions which the new economy of knowledge offers. An example is the trajectory of children and teenagers that join activities such as the Science Clubs of Maloka whereby they can build new paths that they don't find in other scenarios. They develop their talents and interests that will help them to define their vocation and profession while they improve their capabilities for using their leisure time for their own benefit and that of their families.

   The educational programming of the Interactive Centres is an opportunity for renowned scientists, academics, professionals and business people to share their time, experience, and knowledge by means of different communication strategies and become an inspiration and a role model for new generations.

4. *Productive transformation based on innovation and technological development.*

   As engines of an innovative culture, our institutions become "industries of outstanding creativity" thus doing away with the paradigm that culture is limited to art, and that Science, Technology, and Innovation are fundamental elements of our roots and of the growth of humankind. In this sense, the capacity for building a shared vision for a new culture based on learning, knowledge, and innovation is an extraordinary opportunity to transform society in a definitive way.

   Interactive Centres can also constitute links between businessmen, researchers, engineers and academics to bolster their leadership, competitiveness, and entrepreneurial capacity by helping them acquire the necessary tools to channel their creative potential and collaborative spirit.

5. *Political engagement with all leaders for the creation of long-range visionary programs* that include education, science and technology as national priorities for democracy, development and peace.

   As settings of convergence for all citizens, they facilitate the creation of agreements between public and private entities with individuals. Different actors can debate, agree on, and project guidelines for the community in order to respond to local interests and needs. As an example, in Maloka we carry out work with leaders of all sectors on a permanent basis for the purpose of creating awareness about the importance of ST&I for decision making at all levels.

   In the political field, S&T influences the definition of goals, the design of strategies and actions for promoting the use of scientific and technological knowledge in different sectors of society according to its needs, the policies on development and the context in which they will be applied. The *Maloka Political Forum* gathers more than 100 public and private institutions, to promote S&T policies for the Executive and Legislative branches of the government. It also organizes forums and debates aimed at building public opinion on current and complex subjects. We are working for a new law of science and technology to encourage the government to give enough investment and strong policies for research and development.

6. *Articulation of the different "players" in society through inclusive networks*, in order to connect groups with common interests related to development and knowledge, and their place in the daily dynamics of various sectors of society.

   The strategies for the conformation of networks can be oriented to priority groups able to find new channels of communication among them; groups such as teachers, youth, and scientists working alongside the community, enriching their daily work, and significantly improving their social commitment. It is also necessary to think of non-traditional publics such as the Entrepreneurs Network that Maloka promotes, looking for the introduction of scientific and technological knowledge in the productive dynamics of the city.

   The diversity of audiences and allies of an Interactive Centre facilitates the development of networks articulated around common interests. The articulation can be done in different scenarios including the World Wide Web.

   Our renovated Virtual World is designed as a meeting place where dialogue and the exchange of knowledge can help create different strategies aimed at the Social Appropriation of Knowledge and concentrate on information about current science, participation, and the formation of public opinion. It is a multimedia proposal in accordance with a new culture that claims settings where it is possible to share ideas, experiences and tools and where doubts and aspirations can be expressed.

7. *Interactive Centres with a far-reaching range of action, at national and international levels*–enhancement of their coverage and relevance. Our organizations have the capability to increase coverage and reach the most remote regions of our countries with traveling exhibits, educational programming, long distance education, projects with mass media and the creation of networks that articulate pedagogical initiatives with emphasis on Science and Technology. In our case, the work outside the Interactive Centre constitutes one of the major fields of action; through our External Programs we have had a presence in more than forty municipalities around the country by way of strategies such as Traveling Maloka, Maloka goes to School, Maloka and the Stars and the Science and Technology clubs.

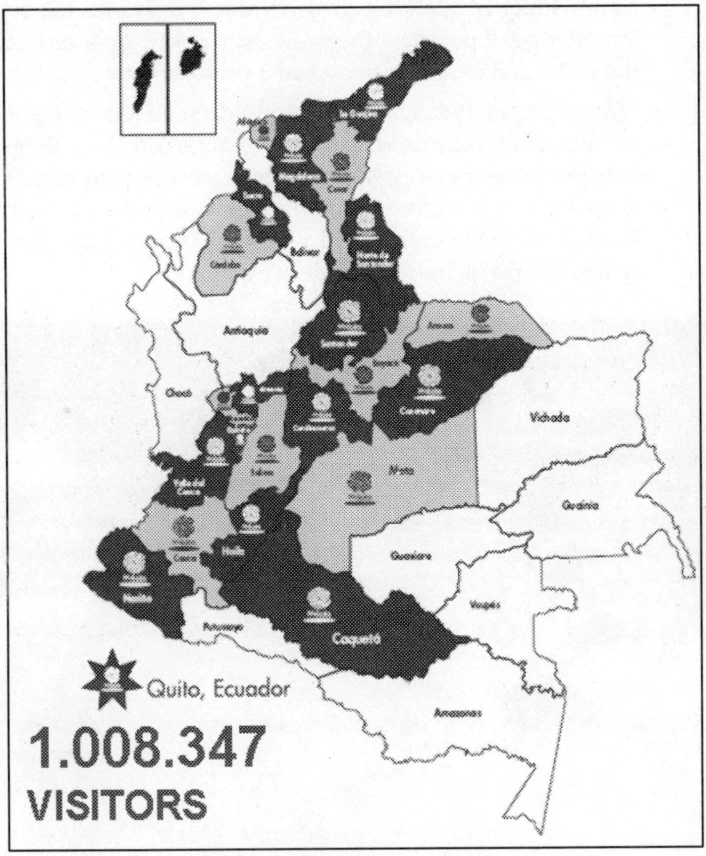

Figure 3.9: Traveling Maloka

8. *Agents of social integration*, as cultural agents entrusted with promoting the development of a scientific culture, our organizations offer different ways of social integration such as:
   (*a*) *Intergenerational encounter*: Adults have the possibility of closing the gap with younger generations. Topics which are so remote and difficult

for older people to grasp are presented in a friendly manner, thus building a new language for the entire family, the most important nucleus of society. Adults can share these new experiences in a simple way, overcoming their fears and their exclusion which are so common in most societies. They are given the opportunity to learn, to use and share new knowledge and its applications and to overcome generational barriers. Senior citizens feel a deepening generational barrier as they consider themselves unable to learn. Often they are excluded from conversations, activities or any task involving the use of new technology. The connection of these senior citizens to our organization is profoundly transforming, because it encourages the refurbishing of life-long projects and it taps into the experience and knowledge of people, otherwise overlooked by society thus allowing the collection of oral history and a projection toward the future.

(b) *Strengthen family ties*, so that the children no longer view their parents as "ignorant" adults who don't understand their language and the new processes, persons who are unaware of reality. In the same manner, the adults of the family can approach its younger members to learn from them and to enthusiastically share the new perspective offered by scientific and technological development.

Figure 3.10: Maloka´s Science and Technology Clubs

Examples of this are the Science and Technology Clubs at Maloka. Children who were hardly seen by their parents because they spent their leisure time at the mall since they were 8 or 10 years old, and now they share knowledge with their parents and grand parents, topics such as astronomy, ecology, or robotics. Our clubs host three generations that have radically transformed their families. At the same time, the children become leaders in the classroom, promoters of innovation both for their teachers and for their classmates.

(c) *Model of inclusion and accessibility*: Interactive centres are non-segregated, non sectarian and accessible to all. The possibility for the most vulnerable groups of society to come to our organizations implies respect for their identity and increases opportunities for all, regardless of their capabilities, age, or physical and mental conditions. The building must have all the facilities needed to welcome the vulnerable and incapacitated. We must be models in architectural design as well as in the added values of its operation and exhibits, offering a readily accessible location for all of the people of the community. The employees in our organizations must be trained to ensure that our settings welcome and help all of our users, despite their physical or mental restrictions. Additionally, our work team must include people with some type of physical restriction who can, at the same time, be leaders in sensitizing our users in this regard. This helps develop talents and provides jobs for those with limited opportunities. The tools needed to grasp, understand and appropriate these topics should be extrapolated as permanent programs (exhibits, teaching materials, books, etc.)

9. *Models of ecological efficiency*

We have a great opportunity for creating organizations that are coherent in terms of the environment:

(a) *Physical infrastructure* which clearly should reflect the innovations of Science and Technology in terms of sustainable development with regard to energy use, water management, logistics and operations, visual and audio contamination, etc. The buildings should remind us constantly, about being active participants about stewardship of the environment and should be replicated in other locations of the country.

(b) *Training and commitment of the employees* with regard to best practices in terms of the environment. In this manner we become an example for the community with regard to of harmonious attitudes and actions to be imitated in everyday living.

(c) *Products and services* for the community developed at Interactive Centres should include topics on the environment and should be articulated transversally with all of the strategic themes that are approached.

(d) Interactive Centres should be *showrooms* for the developments of the scientific and technological community and of the appropriate management of the environment and should also promote those developments that are best adapted to their communities as well as the progress achieved at a world-wide level.

(*e*) The *programs* offered to the community include opportunities for public debates on antagonistic but professional and profound points of view, in which the community and the diverse sectors can participate, bearing in mind that the messages resulting from them need to be adapted to the different audiences. In this way we become coherent, pertinent and influential development hubs that transform the awareness of the citizens and their actions with respect to sustainable development.

Recent Activities

Maloka Political Forum

Since 2006 we started the forum in which we articulate 206 public and private institutions and lots of leaders. We contribute to enrich public policies–Culture, Education, Science, Technology and Innovation. The recent goal of the Maloka Political Forum was the new Law of Science and Technology of Colombia, in which we worked by 3 years. In January of 2009 our President accept the Law before lots of workshops and discussion on our Congress.

We have impacted the executive and legislative power through the construction and enrichment of political documents: local, national and regional development plans, Policy of Science and Technology, Policy of Social Appropriation, Ten-Year Education Plan, Policy of museums, and much more. Besides, we have sensitized public and private leaders on the importance and urgency of these issues to improve the quality of life of citizens.

Network of Innovative Entrepreneurs (NIE)

We have created, in partnership with local government, a network that stimulates innovation on small business entrepreneurs with emphasis on science and technology. Since 2005 the network has served over 700 companies. The results are: 75 per cent improved their marketing strategies. 43 per cent developed plans exporters. 22 per cent increased level of sales by up to 400 per cent . 43 per cent improved its cost structure. 16 per cent reinforced their personal plant. 4.2 per cent (3 companies) made its patent application, which is very significant because in the Colombian industry there are a low number of patents, "less than one patent per year on average."

New Occupations and Ways to Work

It develops talent and interests in young people and adults; aimed at techniques and/or professions based on Science and Technology perspectives. With this program, we will contribute significantly to the transformation of the working capacity of the Colombian people with a vision on new jobs, supporting adults to transform their work and articulate their work in the knowledge society.

Conclusions

Interactive Centres bear great responsibility due to their impact on their community. This impact may be as broad as they want it to be as long as the context is interpreted in an appropriate manner and strong ties of communication and activities with the different segments of society are established.

The strategies developed by these institutions are diverse as are the multiple avenues for community interaction. In this sense, it is important to note that there is no single formula and that the universality of science does not presuppose universality in the ways of appropriating it. This is why it would be pretentious to think of a single model for an Interactive Centre as well as making generalizations about the impact that the Centres may have. Nevertheless, it is necessary to share evaluation experiences and strategies that allow the continuous measurement and improvement in the running of these organizations.

The implementation of networks is of great importance since it affords us the opportunity to share lessons, which will contribute to the achievement of our mission, locally, regionally, and internationally, taking into account the commonalities as well as the differences in our institutions. It is also necessary to identify and characterize the different types of actions that are generated and the dialogue generated with the various populations in each context. Thus to be able to evaluate the pertinence of each strategy and to clarify each question in terms of the relationship that we build with our community and the ways in which we promote the appropriation of S&T as well as the manner in which we participate in the building of identities and diverse cultures based on Knowledge.

Finally, it is imperative to note the relevance of Interactive Centres as venues for integration and social inclusion. In these places, beyond the contact with science and technology, people find possibilities for building identities, for understanding their environment, for developing their personal and professional capabilities as well as opportunities to be with their family and with people from other places and different socio-economic backgrounds. Therefore, the centres become places for the formation of involved citizens able to create and improve the conditions of their environment all of which, in the end, opens the possibilities for building a more just society with better opportunities for all.

References

Colciencias, *Expansion waves*. 2001/2003. Colciencias Wave Program, SENA and Social Fes Foundation, Bogotá, D.C.

Colciencias, *The Colombian perception about Science and Technology*, Bogotá, 2005.

Colciencias–SPE-DCC, Presidency of the Colombia Republic, Social Appropriation of S&t Policy, 2004.

Chaparro Osorio, Fernando. "*Conocimiento, Innovación y construcción de sociedad: una agenda para la Colombia del siglo XXI*". Colciencias, Bogotá, 1998, pp.120

http://www.agenciadenoticias.unal.edu.co/articulos/ciencia_tecnologia/ciencia_tecnologia_20061114

http://www.cab.int.co/cab8/index.php?option=content&task=view&id=674 Cuba, observador de la Primera Semana de la C y T en Colombia.

Interciencia, "*The social Appropriation of Science and Technology: A Regional Need*, INCI v.27 n.2 Caracas Feb. 2002, pp. 53-55.

Revista Ciencia, Tecnología y Sociedad–Número 10–Enero 2008–Apropiación Social de la Ciencia

www.planetariodebogotá,gov.co

www.pequeñoscientificos.uniandes.edu.co

www.cab.int.co/cab1/index.php?option=com_content&task=view&id=47&Itemid=0-23k-

www.musodelosninos.org.co/

www.redpop.org

www. maloka.org

http://www.acac.org.co/apc-aafiles/MemoriasXVIConvencion/27/5 per cent 20HOMENAJE per cent 20MEDALLA per cent 20PROCIENCIA.pdf

Villaveces Cardoso, José Luís. "Los grupos de investigación: bases para la construcción de una política de apoyo a la investigación en Colombia". En; Dinámica y Entorno de los grupos de investigación. Colciencias, Bogotá, 1997.

Chapter 4

Enhancing the Reach through Digitalisation

K.G. Kumar
National Council of Science Museums, Kolkata, India
E-mail: kgkumar@ncsm.gov.in

ABSTRACT

The "scientific literacy" may be defined as "knowing basic principles and concepts of science and understanding how science works for our benefit" (Mukherjee, 2006). Science Museums are in general viewed as an educational institution with a serious role of educating the public. The museums have fundamentally four functions: Collections, Conservation, Documentation and Preservation. In fact now the fifth dimension of the museums has appeared "Education" and it is very obvious that this approach will slowly march towards the sixth dimension "Transformation". The concept of education in museum has been changed from the artifact based mode to an activity based process. The main idea is to incorporate a shift from a 'seeing and reading' mode to a 'doing mode' in learning science. The Exploratorium in San Francisco and some other institutions introduced the 'Science Centre' concept into the non formal science education scenario. The Exploratorium being the pathfinder has influenced many science museums in the world to adopt the 'Doing Science' approach of science centre and to share a common mission–to make science and technology easily accessible to the public and school groups through hands-on exhibits, demonstrations, shows and workshops and many other modes. There are three great elements in science that form the three pillars of modern science in the 20th century–the ATOM, the COMPUTER and the GENES. These three scientific revolutions are dynamic engines for the wealth and prosperity in 21st century. It has become increasingly clear that humanity has the resources to address its global challenges; what is less clear is how much wisdom, good will and intelligence will be focused on these challenges. A science museum would be a right platform for developing a scientific spirit and vigor for facing such challenges by enhancing the reach. We focus our attention on how we will deal with the growth of science centre movement in India (National Council of Science Museums (NCSM), its present status, methodologies by which science centres and museums could organize science education programme using the available technologies, how the science centres and museums can use the digital platform for enhancing their reach and the efforts that are being done in India by NCSM and analysis of the impact of science centres and science museums in developing science literacy in India and other countries.

Keywords: *NCSM, Science communication, Exhibits, Science centres, Science parks, Educational program, Taramandal, Digital, Science fair, Science seminar, Science drama.*

Introduction

Although all science centres have their own slightly different philosophies, they each share a common mission–to make science and technology more accessible to the public and school groups through hands-on exhibits, demonstration shows and workshops (Errington *et al.*, 2001).

If we see the history the Science Centres have evolved from personal collections of curios for curiosity and display of objects for posterity with interpretation and establishment of context to working exhibits-demonstrating process of technology, interactive exhibits to engage visitors, minds-on exhibits to create immersive experience for visitors, virtual exhibits both for in-house and on-line use. In the early days the attitude towards education and career in science was quite favorable to build high-class human resource in scientific fields. The scenario changed gradually and now a situation is created where the job market strongly influences the career of young individuals for joining technological studies just because a technology education would provide them with better means in their life than pursuing higher studies in science. There is a wide gap in the formal science education system between the theoretical aspects and the practical demonstrations in the classes. Science education has now strongly become a chalk and talk activity in the formal science learning system. It is in this context, the science centre activities can play a very strong role to arouse interest and curiosity in science among the young minds. The 'Doing Science' approach of science centres in fact helps students in gaining a deep-rooted concept related to various scientific principles. No doubt, this builds up an affinity and interest in science among students.

From the period 1960 onwards, a new concept of popular science exhibit was conceived to introduce the excitement in museum learning. In this learning atmosphere, visitors from different backgrounds were offered a scope of doing science. Each of the new interactive exhibits was designed to convey some message or principle of science through a two-way enjoyable communication between the exhibit and the user. During this changing scenario in the museum field in the world, the Planning Commission of the Government of India, with a view to create scientific awareness and temper in the common man for the benefit of the society, created an auonomous society-'The National Council of Science Museums'(NCSM) on April 4, 1978.

The growth of science centres in India, under the control of NCSM was distributed among National, Regional and District levels with different sizes and scopes. The development of science centres of such categories created different technological environments associated with the type of exhibits according to the availability with the local resources for maintenance and upgradation. The scope of using sophisticated techniques was thus constrained by the location and category of science centres and consequently a challenge (Kumar, 2006) in more diversified technical approach in exhibit making was opened up. An innovative measure was thus required to meet those challenges.

This challenge was taken up by NCSM and subsequently the Council has excelled in its capability to produce state of the art exhibits, new communicaton and display techniques and interactive exhibits.

NCSM is now working with a large network of 27 science centres. This is the largest network in the world under a single administrative control. The Council has earned attention and recognition from all parts of the world for its widespread innovative technology-based activities in science communication.

Fundamental Digital Concepts

We have entered a digital era. Evidence can be seen everywhere. Computers out-number office personnel, e-mail addresses as well as web pages are certainly taking the role of a tool for individual identification. Besides, now it is no longer essential to carry cash for business transactions there are computers and satellites to help us.

At the most fundamental level, digital technology is an extension of the long history of the way we communicated with each other (Ray, 2006). The desire to communicate provides motive and the ultimate rationale for the development of technologies of all sources. Today's digital world is concerned with creating, sharing and using information in digital form.

Creation of a Culture of Science is a Primary Necessity for National Development

Science Communication is an essential requirement for creating a culture of science (Mukherjee, 2007).

Digital Electronics has grown rapidly in the domain of information and communication.

Information and communication technology (ICT) has become extremely important for communicating science in an effective way to a broad spectrum of target audiences.

ICT is going to be a very major tool for achieving the objectives of science museums and science centres for

Supplementing 3-D exhibits in a science museums/science centres.

Providing CD-ROM based educational tools.

Creating high-class animations.

Creating virtual reality.

Customizing exhibits through image processing techniques.

Communicating science through satellite channels.

Communicating science through Internet.

Evaluation through digital techniques.

Digitization

One of the questions that we like to address is how can the science centres ensure that the opportunities offered by digitization are fully utilized for the benefit of visitors to the centre as well as to those who are not visiting?

There is an urgent need for a planned approach to the Digitization of collection with the available digital framework to provide strategic content creation and a platform for easy public access to the collections of the museum. The outcomes of digital archiving are as follows:

- Fragile artifacts could be displayed with no further risk of deterioration.
- Artifacts could be shared among researchers and the public with no limitations on distance, transport or exclusivity.
- Artifacts could be displayed in ways that enhance their contextual relationships with the environment in which they were originally created and used with no risk to the original.

Science centres have the potential to make a significant contribution to e- Learning (Kumar, 2007).

Basically a Web-based learning system can create an environment that will bring together effective informal learning strategies and formal curricular activities in a single platform using digital techniques. A network of science museums/centres, research institutes and schools can be established to provide an advanced learning environment. This will incorporate innovative use of the technology for educational purpose and will also interconnect all the members of the network.

Further it will also organize virtual and conventional visits of the students to the science museums/centres. These visits will support (through an informal but yet structured way) the main educational aims of the curriculum. The aim of this is to explore, test, refine and demonstrate an innovative approach that crosscuts the boundaries between schools, research institutes (*e.g.* observatories), and science centres and involves students and teachers in extended episodes of playful learning.

A well structured and well *organized website* (Mukherjee, 2007) will help the visitors to explore the available scientific information in the science centre. In addition to offering 'Virtual Access' to a science centre galleries and artifacts to remote audience the archives can also be shown in this platform. The capacity of web browsing software to integrate text, images and sounds has attracted many science centres to the idea of establishing a presence on the web. Significant developments in web browsing software now make it possible for web users to experience far greater levels of

Figure 4.1: An Exhibit on 'Virtual Experiments in Physics'

'interactivity'. Therefore, science centres around the world are inviting people of all ages to enter the fascinating world of science and technology through their computer. The advanced web technologies enable the online visitors to interact with sophisticated animated sequences and simulations, navigate in three-dimensional environments etc. (www.ncsm.gov.in)

Challenges–Translating Physical Exhibits in the Digital Format

Physical exhibits in a science centre require a significantly large investment of human capital as far as the expertise and content are concerned. Physical exhibits *project* the identity and vocabulary of the educational framework of the science centre whereas the virtual exhibits cannot precisely perform as physical exhibits. It would be of interest if we evaluate how some of the different types of exhibits in a physical science centre can or cannot be readily adapted in the virtual domain.

Although there are many ways to categorize exhibits; this treatment is not exhaustive.

The Table 4.1 lists several science centre exhibit types and their possible digital equivalents.

Table 4.1: Science Museum Exhibit Types and their Digital Potential

Physical Exhibit Type	*Digital Potential*
Artifacts based exhibit	Objects become part of reference encyclopedia and the digital medium can talk about the history and details of the objects with animations.
Demonstration of a physical principle	The underlying principle can be explained in details instead of demonstrating surface phenomenon, as may be the case.
Illustration of function of human cognition	Can be equally effective.
Mechanical analogy for non-mechanical principle	Find new non-mechanical analogies and the actual phenomenon can also be shown in animation or in real environment.
Fun science or hardware-derived tricks based	New digitally origin opportunities can be introduced for interactive play.
Showing exotic and emerging technologies	Can explain underlying principles and innovation in the field with more interactivity and related scientific phenomenon leading to the development of the technology. Can also initiate a debate on the myth and realities of the technology and related socio economic developments or otherwise.
Exhibits on socially relevant topics and issues	Provides a strong opportunity to frame and initi ıte community discussion.

Digitization increases our capacity to *reach new audience* nationally and internationally at a time when some centres are reaching the limit of the number of physical visitors they can accommodate. Prior to the early 1970's specifically in India, access was primarily through on-site tours. People had to travel to the location of the centre and once inside move through the various display and exhibition spaces before the closing time. Traditional science centres included various artifact displays, exhibits and hands-on experiments with the usual restrictions of space and hours of opening. In an attempt to overcome the access limitation of 'on-site only access' Mobile Science Exhibitions and later travelling exhibitions and displays were introduced. For the science centres such programs expanded the annual visitors from hundreds to millions (Andover, 2004).

Whenever a visitor interacts with an exhibit, talks to an explainer, listens to a public lecture, watches a science demonstration show or participates in a hands-on activity session, the experience can only take place when the visitor is actually there at the time the experience is scheduled (same place–same time).

Whenever a visitor connects himself to a science centre web site and interacts with an online activity, reads information, listens to a sound recording, observes a video clip or enters a response as part of an online competition, the experience can take place when the visitor is remote from the science centre and is not restricted by the opening hours or program schedules of the centre.

By providing experiences for online 'visitors' as well as actual visitors, it can be argued that science centres may reach and influence many more people. The purpose of science centres is more than getting a greater number of visitors through their actual or 'virtual' doors. The cyberspace is slowly becoming an annex of traditional science centres.

Video conferencing is another avenue that enables the visitors all over the country to access and explore the resources of the science centre. Right management is an issue for the digitization project and can require significant input of resources. In line with this to provide more access to the remote visitors and with the aim of simultaneously transferring information from different parts of the country, NCSM had a vision to have *connectivity through V Sat* enabling National level centres, the Headquarters and the Mobile Science exhibition users to interact in science programmes. This has been achieved partly by now.

Figure 4.2: V-Sat Communication Network Set up at NCSM Headquarters

Scope

With the easy availability of Broad Band facility in our country and the rapid evolution of free or low cost multimedia capable browsers we should plan to move into Cyberspace. This can offer a range of educational and "hands-on" learning materials to a global audience.

As the *mobile Internet* use is growing so is the expectations of the audience in receiving information anywhere at any time. The opportunities of the mobile Internet to deliver information to the users within appropriate content is to be explored in our country.

The visitors to science centres bring with them great diversity in terms of their previous knowledge and experience, their assumptions and their expectations, as well as their ways of thinking and learning. Science centres expect this diversity and design their exhibits and programs to the extent possible to take care of the diverse visitors needs. In the same fashion the online exhibits and activities for virtual visitors should also be designed keeping this principle in mind.

When actual visitors interact with hands-on exhibits and activities in science centres, they encounter science phenomena in engaging ways, manipulating devices and data to test and develop their understanding-the more compelling the experiences, the more likely learners will develop their understanding of the phenomena. Virtual visitors, too, are more likely to develop their understanding if online activities are designed to enable people to test their ideas.

New Technology and NCSM's Science Museums/Centres

In NCSM science museums/centres, mainly through the R&D endeavours, efforts are on to develop exhibits through applications of new technology and cater requirement of visitors who look for an experience that is of immediate personal relevance and which ensures a clearly identified knowledge gain. This paper represents some applications of the new technologies in indigenous development of exhibits by NCSM during the last couple of years.

Virtual Chemistry Laboratory

Setting up chemistry or a physics lab can be a costly and time-consuming endeavour for the schools. The inconvenience becomes even more pronounced when institution happens to be a remote one or a rural one. An alternative solution of the problem has been attempted through development of interactive multimedia programmes, which presents an exact environment of a science lab on the computer screen. The user can select tools and ingredients and virtually perform the actual experiment and see the result of the same. The virtual chemistry lab has been developed and the physics and biology lab is under construction. NCSM has obtained a Copyright for this software.

Interactive Planetary Globe Exploration System (High Class Animations)

Virtual Planet–Hemispherical Projection

A dynamic projection of animated computer graphics of a planet on a hemispherical screen may results into a very realistic dynamic model of the same provided the 3D object wrapping maps are generated in high resolutions. Once the models are made, suitable software tools can be developed, which allow the user to manipulate and manoeuvre the 3D objects through a multimedia interface.

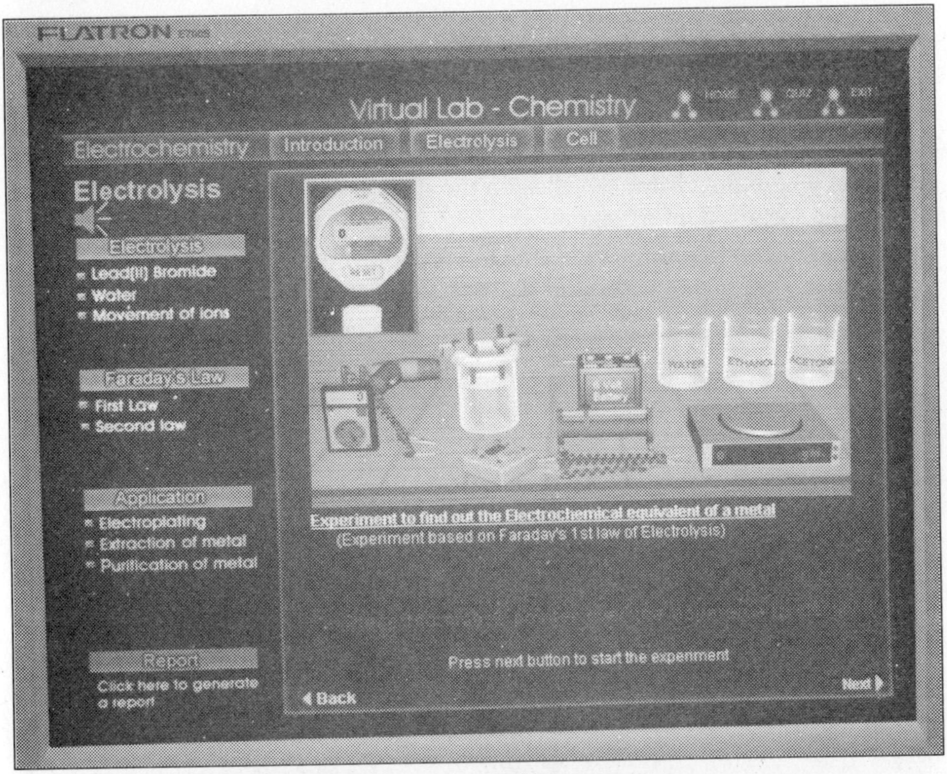

Figure 4.3: An Interactive Computer Based 'Virtual Laboratory' on Chemistry

Figure 4.4: Virtual Planet Exhibit Using Hemispherical Projections

This enables the visitor to see any planet in various surface textures–topographic, bathymetric, political, riparian features, geological plates, volcanic belts, satellite view etc.

These versions had only the first level interaction where the user can change the planet and grab and rotate it. Work is now on for providing further feature sensitive interaction where the visitor can click on the surface of the globe and extract more relevant and localized information.

Technology for fabrication of such exhibits is now fully realized the real challenge lies in appropriate and rich content creation on which also we are working.

Figure 4.5: Virtual Planet Exhibit Using Hemispherical Projections

New Display Technique: The Unmanned Quiz Master

Museums of the world continuously search for a method to take visitor feedback with respect to evaluation of their understanding of the exposition. Quiz works well in this regard. But all quiz programmes in their familiar form require a quizmaster with the participants and a scoring system. Interaction of the quizmaster also becomes essential for the decision making about the correctness and ambiguity of answers. Such quiz shows can only be held at definite time slots. To avoid human intervention and to allow maximum interactivity of the visitors we thought of a new presentation technique that can be operated by the visitor without the human intervention.

Advantages of the system are possibility of random question selection, immediate feedback analysis and presentation of relevant audio and video graphic and information along with the quiz sessions. A system has been developed by NCSM

Figure 4.6: An Unmanned Quiz on "Oil"

and now the technology has been furthered that it can be tailor made for the required number of audience and in any desired language. Several systems have been installed and they are running in public environment of the museum galleries with great success. The system was developed in CRTL and Concept Note Copyright for this exhibit has been awarded to NCSM.

Context Pickup Mode Information Retrieval System (Moving TFT/ Visual Seeker)

Historical timeline presentations are of utmost importance to the museums of the world.

Traditionally such presentations are accomplished either through static graphical displays, or in computer kiosks. A typical problem while presenting these information is that either it is inadequate or too much and the visitors cannot customize the retrieval of information as they wish.

Looking for a solution a system has been developed where the visitor can move a flat TFT display in front of the 2D timeline presentation in a visual form. There are certain nodal points in the 2D timeline display; once the visitor moves the TFT display on those point further details of that time zone appear in the TFT panel. Visitor can explore more information on related topics by going deeper in the software through hyperlinks shown in the TFT panel by touch sensitive system.

This system has been fully developed and a patent for this technology has also been done by NCSM.

Figure 4.7: An Exhibit Using the 'Visual Seeker System'

3D Scientific Documentary Video Presentations using Anaglyph Technology

A normal video projection presents two-dimensional views the viewers. The third dimension can be virtually imparted to the viewer if two perspective views of the same scene can be rendered separately in two eyes. One technique to do so is to use anaglyphs for each frame in the video. Suitable technology can be built around 3D model creation software tools, virtual cameras and anaglyph software tools, which enables development of the 3D content. Taking stereoscopic live video requires a costly stereoscopic video camera system. But an alternative is envisaged in creating an animated world through 3D model creation software tools and rendering two perspective views by placing virtual cameras in the scene. These two separate views are then recast into red and blue channels and juxtaposed in the same projection frame through re-rendering by anaglyph software tool. The advantage of such development is that 3D documentaries appropriate for the theme of the galleries that are planned for the Science Centres can now be developed in house by NCSM. Central Research and Training Laboratory (CRTL) indigenously developed the technology for developing 3D Scientific documentary and the first such movie named 'The Group of Nine' was screened at Kurukshetra Panorama and Science Centre (KPSC) and in Illusion Exhibition. Rajiv Gandhi Science Centre (RGSC) runs a permanent show of the movie.

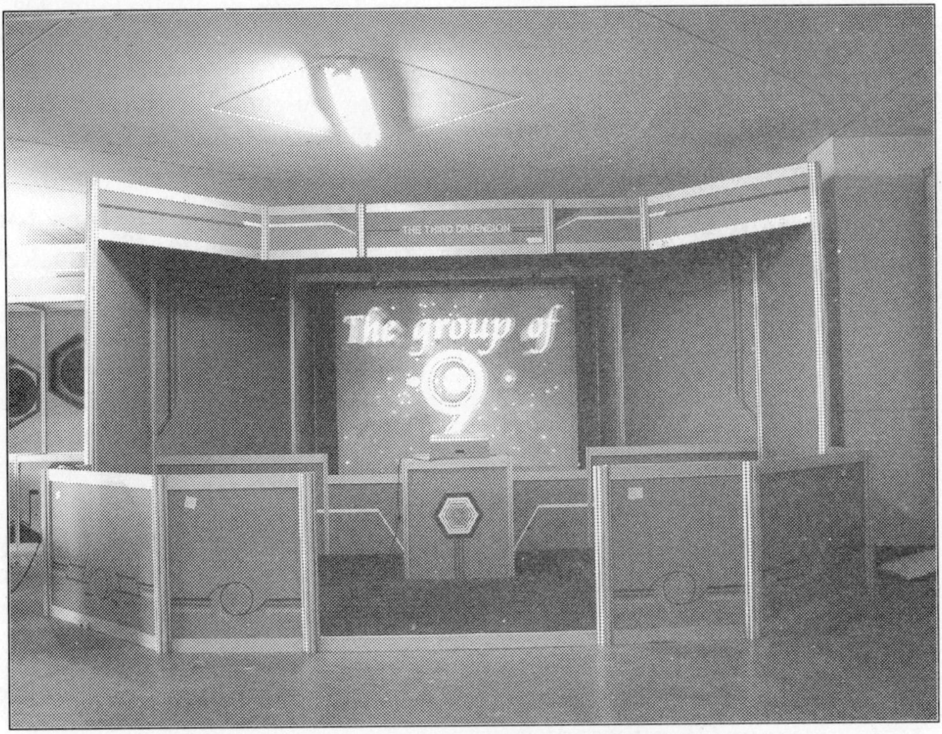

Figure 4.8: A 3D Anaglyph Video Projection System

3D Scientific Documentary Video Presentations using Dual Electronic Output and Polarizing Filters

The intention for taking up this project was to develop a technology solution to produce low budget scientific movies having virtual depth perception in the third dimension using the dual field separation embedded into video frames. Final viewing

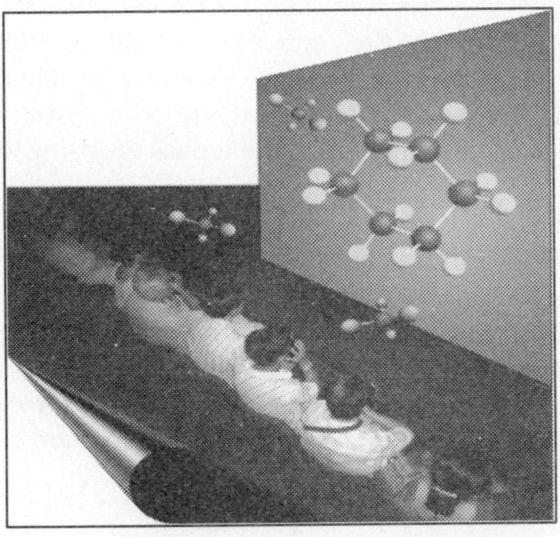

Figure 4.9: Viewing a 3D Movie

would be through Polaroid films. This was really a development over the technology achieved during the creation of anaglyph movie. The context was chosen from the requirement of one such video presentation necessary to be displayed in the Oil and Natural Gas Corporation (ONGC) Golden Jubilee Museum. The story–Precious Molecules–obtained significant applause from the viewers. It is running with success and popularity in the same museum.

Development of Hardware Interface and Software Engines for Interactive Gaming Consoles Used in Science Dissemination (Basket Ball Game)

Science Centres quite often use pre-rendered animations and movies to present certain ideas of science. However, it is seen that interactive modes of presentations work better when the user is given to handle an interface familiar to him/her. It is even experienced that hiding the PC/Controller to act from behind and taking inputs from common mechanical interfaces (like steering, lever handles etc.) keeps the user far more interested to work with the exhibit.

So, a core exhibit was planned which would implement Computer Based Training units via symbiosis of PC and common hardware interface elements. These required online rendering facilities for real-time presentation of scenes involving multiple user perspectives. A first level prototype for this exhibit has been developed.

Non-Linear Frame Rate Variation of Video Graphic Files (Seeing Video at your Own Speed)

Motion picture has always been a strong medium of science communication. But the normal speed video, while using the principle of persistence of vision, hides a lot of motion details from the comprehension of the viewer. For example, a slow motion video of a bird's flight or a speed-enhanced movie of a blooming flower may convey many details of the corresponding transformation mechanism. This technology is envisaged to transfer the motion or frame rate control of any video to the hands of the visitor, such that, the visitor can playback the video, through a simple to use interface, at their own chosen speed/rate. CRTL has developed suitable live video capturing mechanism from the pre-controlled and guided interface appropriate for the museum, and developed the API-level video frame rate control and playback tools in software. The next step in this regard is to develop a suitable visitor interface and to develop a complete deliverable exhibit package using this technology.

Interface Design–Haptic Interfaces

Haptic technology refers to technology that interfaces the user via the sense of touch by applying forces, vibrations and/or motions to the user. This mechanical stimulation is used to create haptic virtual objects. This emerging technology promises to have wide reaching applications in science centre exhibits. NCSM is engaged in developing several such exhibit interfaces which will not force the visitor to adapt to any special skill (like controlling a mouse pointer) to interact with a virtual space exhibit.

Figure 4.10: Development of the Non-linear Frame Rate Variation of Video System

Figure 4.11: Visitors Interacting with a New Exhibit 'Treasure Hunt'

CRTL has developed one walk-on exhibit interface with floor mounted pressure sensitive electronic control that allows the visitor to control a computer game with his gesture and full-body movement. The system can be used in developing several types of museum applications like mathematical game simulations. In the first version of this exhibit, the visitor can walk on a camouflaged floor sensitive to the pressure and that can determine his/her position on the floor to and within the limit of a resolution of 1 ft x 1ft. Using the Walk-on interface, developed by us the game Treasure Hunt for which a Trade mark application has already been filed and the patent is being applied.

Dome Projection

NCSM has developed the technology of creating content for the immersive dome projections. Importing the Fisheye Lens projector and assembling the unit with the help of all indigenously developed software and dome; NCSM has been able to fabricate a digital planetarium system and a dome video content projection system. NCSM is employing this technology to install 80-seater planetarium in three Regional Science Centres.

2D Space Browser

NCSM has developed a document browser technology wherein a large high-resolution digital document is projected on a table that can tilt on a universal

Figure 4.12: A Taramandal Demonstration

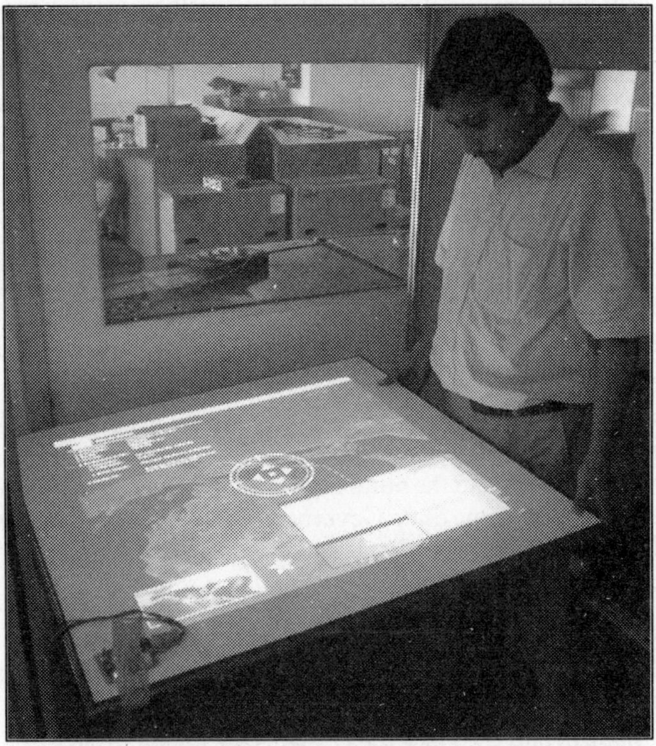

Figure 4.13: An Exhibit 'Map Browser' as Outcome of Research and Development

mechanical joint and can be tilted at any angle. While tilting the projected digital document scrolls synchronously. Changing the tilt angle varies the document scrolling speed. By this system the visitor can pinpoint the required information in the document by controlling his scroll speed.

To develop the technology required software tools were not readily available. So, they were indigenously developed by NCSM and this generic tool has now been extensively used in many other exhibits.

Though similar exhibits are fabricated by Japanese Company, Onomy Labs, NCSM is preparing to file a patent claim on this technology by representing the novelty claim that it has developed a low cost solution based on software RS-232 communication tools. This technology may in future be used to present contextualized maps of India that do not exist in geography domain. For example, it may be used to present an anthropological profile of India. Content for such exotic maps can be prepared in collaboration with other related organizations like ASI etc.

Conclusion

As we continue to push the boundaries of the digital frontier, we are also redefining the scope and range of our mission as public learning facilities. By implementing the latest advances in digital technologies with wisdom and vision, the students as well as common people can team up in a unique way for understanding new scientific advancements (National Research Council, 1996; National Science Board, Science and Engineering Indicators, 1998; Ansbacher, 1998).

In the Indian context and with the available technologies there are certain questions that require to be addressed.

Are we fully connected to the users? The answer is partly true, because save the metro cities and the capital cities the broadband network is gradually reaching other places. Even though we have the Broad Band Network, more connectivity speed is required. We at the science centre end may be having an efficient system but the end user system should also have the capacity to handle the large data that is made available by us in the web or in the digital medium.

Similarly many schools even in the metros are yet to be networked to each other although efforts have been initiated in some cities for having leased line networking with schools–but once again the number of schools is less. What can we do about the Government schools? Some State Government have taken an initiative to provide connectivity to the schools–but that is still in a stage of infancy.

Therefore, can it be that the science centres establish the contact with the

1. Government and convince the authorities about the viability and usefulness of the network system so that an initiative is taken and at least a pilot project is started in each state?
2. Establish and convince the IT giants for taking up this cause, jointly with NCSM?

NCSM by itself decided to initiate some activities in this area that have been outlined above.

The second step that is envisaged is to network through satellite connectivity the major science centres in each zone and the Headquarters of NCSM–the networking will be also extended to the Mobile Science Exhibition (MSE) units in each zone. This will allow all the major units of NCSM to share their programmes and interact with the activities on line. The visitor reach will be increased due to the same network available at the MSE sites. However these programmes and activities will still allow us to reach only those visitors who will come to the museum or to the MSE sites.

As a third step–NCSM and its units are in the process of launching the web based activities, since by the time the connectivity speed of the network system would increase the science centres should also be ready to deliver the content.

Exhibits based on application of digital technology have gradually become *a just inclusion* in the galleries of our science museums/centres. It can be stated with certainty that the digital technology is here to stay and our valued visitors expect to be able to interact with such exhibits.

Technology is effecting the education in revolutionary ways and the *momentum* towards these changes is *irreversible*. Various capabilities achieved as an outcome of the R&D endeavour of NCSM have not yet become functional in a big way, *but they will be very soon.*

With today's technological revolution as infrastructure, all sorts of novel activities can occur and people are becoming increasingly used to the technology. Therefore, a time has come, when the museum professionals should be aware that people's expectations of what the technology is capable of are increasing. But it is to be emphasized that the content of a science museum is the most important part and we may conclude by saying that the content is all and we should not be sucked into the technology for its own sake. Technology should be used to enhance the visitor's experience but technology should not dictate the experience. I hope that the science centres of the 21st century will find the courage and vision to experiment with different styles and techniques to engage new audiences and deal with a wider range of issues. Like science itself, their future lies in continual innovation and evolution (Barrialault, 1998; Falk and Dickering, 1992; Falk *et al.*, 1998).

References

Andover, MA, 2004. Digital Technology made simpler by Conway. P North East Document Conservation Centre.

Ansbacher, T, 1998, John Dewey's Experience and Education, Lessons for Museums, Curator 41, pp.36-49.

Barrialault, C, 1998, The Science Centre Learning Experience: A Visitor Based Frame work. M.Sc. Thesis, Glamorgan University, Techniquest, Cardiff, Wales.

Errington. S. *et al.* (*Eds*), 2001. Using Museum to Popularize Science and Technology A Commonwealth Secretariat Publication.

Falk, J.H and Dickering, L.D., 1992. The Museum Experience, Washington DC, Whalesback Books.

Falk, J.H, Moussouri, T. and Coulson, D., 1998. The Effect of Visitors agendas on Museum learning, Curator 41, pp.106-20.

Kumar, K G, 2006. 'Facing the Challenges of New technologies in Science Centres', *In* Proceedings of the 5th Annual Conference of the heads of Science Museums and Science Centres held at the Regional Science Centre, Bhopal, India, 10th to 12th Feb. 2006.

Kumar, K G, 2007. 'Internet in Science Communication', *In* Proceedings of the 7th Annual Conference of the Heads of Science Museums and Science Centres held at the Regional Science City, 7th to 9th Dec 2007, Lucknow, India.

Mukherjee, I K (*Ed*), 2006. Reinventing Ourselves–New Models for Science Centres in India, *In* Proceedings of the 5th Annual Conference of the heads of Science Museums and Science Centres held at the Regional Science Centre, Bhopal, India, 10th to 12th Feb. 2006.

Mukherjee, I K, 'Science Communication on the Digital Platform', *In* Proceedings of the 7th Annual Conference of the Heads of Science Museums and Science Centres held at the Regional Science City, 7th to 9th Dec 2007, Lucknow, India.

National Research Council, 1996, National Science Education Standard, Washington D.C., National Academy Press.

National Science Board, Science and Engineering Indicators-1998, NSB 98-I, Arlington, VA, National Science Foundation.

Ray, S K, 2006. 'Application of New Technology in Science Museums/Centres of NCSM- Augmenting the learning experience', *In* Proceedings of the 5th Annual Conference of the heads of Science Museums and Science Centres held at the Regional Science Centre, Bhopal, India, 10th to 12th Feb. 2006.

Glossary

NCSM: National Council of Science Museums
CRTL: Central Research and Training Laboratory
R&D: Research and Development
ICT: Information and Communication Technology
KPSC: Kurukshetra Panorama and Science Centre
RGSC: Rajiv Gandhi Science Centre
ONGC: Oil and Natural Gas Corporation
MSE: Mobile Science Exhibition

Chapter 5

Role of Pushpa Gujral Science City for Sustainable Development

R.S. Khandpur
Pushpa Gujral Science City, India
E-mail: rskhandpur@rediffmail.com

ABSTRACT

Pushpa Gujral Science City (PGSC) at Kapurthala in North India has been set up for supplementing science education, widening public knowledge, providing avenues for innovation, guiding appropriate career selection, demonstrating emerging technologies and national strength in the field of science and technology. The objectives are being fulfilled through numerous areas of activities by providing hands-on and minds-on experiences with the help of interactive exhibits and linking the exhibits to modern school curricula to reinforce the formal learning experience. The Science City is playing an effective role for creating awareness about issues related to sustainable development. With this in view, PGSC is attempting to focus on areas of direct concern to human development, such as energy, health, biodiversity, biotechnology, rain harvesting, waste management, agriculture, transport and environmental challenges. With the edutainment approach followed in the Science City, the public is being sensitized about the delicate balance that exists between the present day human needs and life style, and preserving natural resources and echo systems. The Science City is effectively disseminating information on ways and means to meet the needs of the present without compromising the ability of the coming generations to sustain them. PGSC is playing a pivotal role in spreading awareness about measures that can contribute to combat global warming. Demonstrations and special programs have been developed to educate the visiting public in popularizing the renewal energy sources, conservation of water by use of sprinklers or drip-irrigation devices for watering the lawns, recycling of waste, popularizing afforestation, saving energy and reducing air pollution etc. Special films on the effects of climate change are regularly screened.

Keywords: *Sustainable development, Pushpa Gujral Science City, Science centres, Science museums, Science park.*

1. Introduction

Although tradition of Science and Technology (S&T) in India is over 5,000 years old, the real push came only after India became an independent nation in 1947. The network of national laboratories under CSIR (Council of Scientific Industrial Research) was expanded. Research laboratories were set up under DRDO (Defence Research and Development Organization), ICAR (Indian Council for Agriculture Research) and ICMR (Indian Council for Medical Research). Establishment of the Bhabha Atomic Research Centre and later on, the Indian Space Research Organization has given India a prominent place among the elite scientific nations of the world.

The Government of India is committed to making Science and Technology (S&T) an integral part of the socio-economic development of the country. It has been brought into the mainstream of economic planning in the sectors of agriculture, industry and services. The emphasis is on using science and technology to harness country's resources to derive the maximum output for the benefit of society and improvement in the quality of life.

2. Science and Technology Policy of the Government of India

Considering that the Science Centres can play a very important role in disseminating the message of science to the citizens of India, the S&T Policy of the Government of India announced in 2003 that, inter alia, the message of science must reach every citizen of India, man and woman, young and old, so as to advance scientific temper, emerge as a progressive and enlightened society, and make it possible for the society to participate fully in the development of science and technology and its application for human welfare. The objective is to fully integrate science and technology into all spheres of national activity. While formulating the strategy for achieving public awareness, the Policy states that:

- There is a growing need to enhance public awareness of the importance of science and technology in everyday life, and the directions where science and technology is taking us. People must be able to consider the implications of emerging science and technology options in areas which impinge directly upon their lives, including the ethical and moral, legal, social and economic aspects;
- Support for wide dissemination of scientific knowledge, through science museums, planetaria, botanical gardens and the like, will be enhanced;
- Every effort will be made to convey to the young the excitement in scientific and technological advances and to instill scientific temper in the population at large.

It implies that the activities of science museums and science centres have been identified as an area of great importance for promoting public awareness of science and technology, with the Government of India giving a special thrust to this area (Sinha, 2005).

3. Establishment of Science Cities in India: A New Concept in Science Centres

During the mid-90's, a new concept got introduced in India in the field of science centres. It was felt that for making a science centre truly a visitor's place, the centres must offer much more than exhibitory in the closed buildings. A major initiative in this direction was taken with the setting up of the Science City in Kolkata in 1997. With the popularity and success of the Kolkata Science City having been established, work on setting up a new Science City at Kapurthala named as Pushpa Gujral Science City was started in the late 90s.

As compared to Science Centres which non-formally teach different principles of science and technology through hands-on interactive exhibits, the Science Cities have much broader scope of activities and facilities. These include: a science exploration facility for displaying hands-on and minds-on exhibits in various fields of science and technology; a large format film projection system; space flight simulator; a virtual reality 3D show; a science park and recreational areas like cafeteria, souvenir shop, boating etc.

3.1 Main Objectives of the Science Cities

- To portray the growth of science and technology and their application in industry, human welfare and environment, with a view to develop scientific attitude and temper and to create, inculcate and sustain a general awareness amongst the people.
- To promote and enhance public understanding of the culture of science and technology
- To popularize science and technology in cities, urban and rural areas for the benefit of students and for the people at large by organizing exhibitions, seminars, popular lectures, science camps etc.
- To supplement formal Science and Technology education through demonstrations, lectures hands on exhibits.
- To organise outreach programmes on Science and Technology to reach the people at large
- To actively engage in building up scientific manpower
- To generate a scientific culture in the present day knowledge-based society.

At present, three Science Cities are in operation. These are located at Kolkata, Kapurthala and Ahmedabad.

4. Pushpa Gujral Science City, Kapurthala, Punjab

4.1 Background

Pushpa Gujral Science City (PGSC) has been set up on 72 acres of land on Jalandhar-Kapurthala state highway. PGSC Complex is adjoins the National Institute of Renewable Energy and Punjab Technical University (PTU). This project is a joint venture of Government of India and Government of Punjab, being set up with an investment of Rs.100 crores (US$ 25 Millions).

PGSC has been developed as an institution of excellence–a complex to hum with science activities all the year round and to generate and nurture creativity in young minds. The vision for the Science City goes beyond a museum or a theme park. It aims to facilitate education through entertainment and showcases contemporary and imaginative exhibits, hands-on experiences, working models, virtual reality, activity corners, and labs and live demonstrations to provide an understanding of science and technology to the students and equally well to the common man.

4.2 Mission

Pushpa Gujral Science City has been developed as:

- A place for education: to provide young people with the opportunity to explore informal ways of acquiring knowledge.
- A place for widening knowledge: to acquaint everyone–children and adults, general public and researchers–with the information resources on varied subjects.
- A place for innovation: to provide opportunities to individuals to try out their ideas in the design and development of new products.
- A place for information: to compile and disseminate information on the latest developments in science and technology.
- A place for career guidance: to assist young people to select the most appropriate career in science and technology from a wide spectrum of opportunities.
- A place for demonstrating emerging technologies: such as nano-technology, micro machines, levitations, genomics etc.
- A place to demonstrate the industrial strength of India: through models and visuals of various prominent industries.
- A place to display the national strength in the field of science and technology: demonstrating the developments in its various national research laboratories.
- A place to participate in adventures sports: such as rock climbing, aero-ballooning, boating etc.
- A place of debate: to direct its efforts towards bringing together researchers and participants who are eager to discuss important questions raised by scientific and technological progress.
- A place for career guidance: to act as career guidance centre and hosting exhibitions and fairs on different career sectors and above al!
- A place for edutainment: to provide opportunities for learning through experience and entertainment.

4.3 Strategy

In order to fulfill the mission objectives, the Science City is engaged on numerous areas of activities. The main attractions in the Science City are the exhibits displayed in an interesting manner and so selected that they:

- Reinforce the formal learning experience;
- Relate science exhibit activities to modern school curricula;
- Encourage visitors to learn basic scientific principles;
- Serve as a catalyst for improved science education;
- Provide hands-on and minds-on experience and
- Are interesting and exciting for visitors of all ages.

5. Science City Exhibits

The Science City has been planned to have both indoor and outdoor interactive exhibits spread over several galleries. A layout plan of the Science City is shown in Figure 5.1. The details of the galleries and other facilities set-up in the Science City are as follows:

5.1 Major Attractions

- Space Theatre in which the Large Format films are projected on 23 meter tilted dome to provide the visitors immersible experience. Figure 5.2 shows the exterior of the Dome Theatre along with the Aviation and Space Gallery Building.
    - *Flight Simulator* provides visitors a sense of adventure through virtual expeditions to moon and centre of the earth.
    - *3-D Show*: Virtual Reality show on different themes of science and technology.
    - *Indoor Laser Show*: displays spectacular graphics and beam shows on popular subjects.
    - *Earthquake Simulator*: provides real life experience of earthquake of different intensities on the Richter scale.

5.2 Galleries

- Aviation and Space Gallery displays cut-view models of satellites, Robotic planetary probes and various applications of space technology, hands on exhibits in the field of aviation and astronomy.
- Energy Park displays applications of different types of non-conventional energy sources. Science City has established a Solar Energy Power Centre, Hydel Energy Centre, Bio-mass Energy Centre and a Solar Restaurant.
- Health and Biotechnology focuses on spreading awareness about HIV/ AIDS. The Gallery also shows anatomical and physiological functioning of the various organs of the human body. It has special sections on HIV/ AIDS and Bio-technology applications in the health Sector.
- Evolution Dinosaur Park has been set-up on the island surrounded by artificial lake which provides a magnificent view.
- Fun Science enables the visitors to understand the basic laws of Physics in an interactive and playful manner.

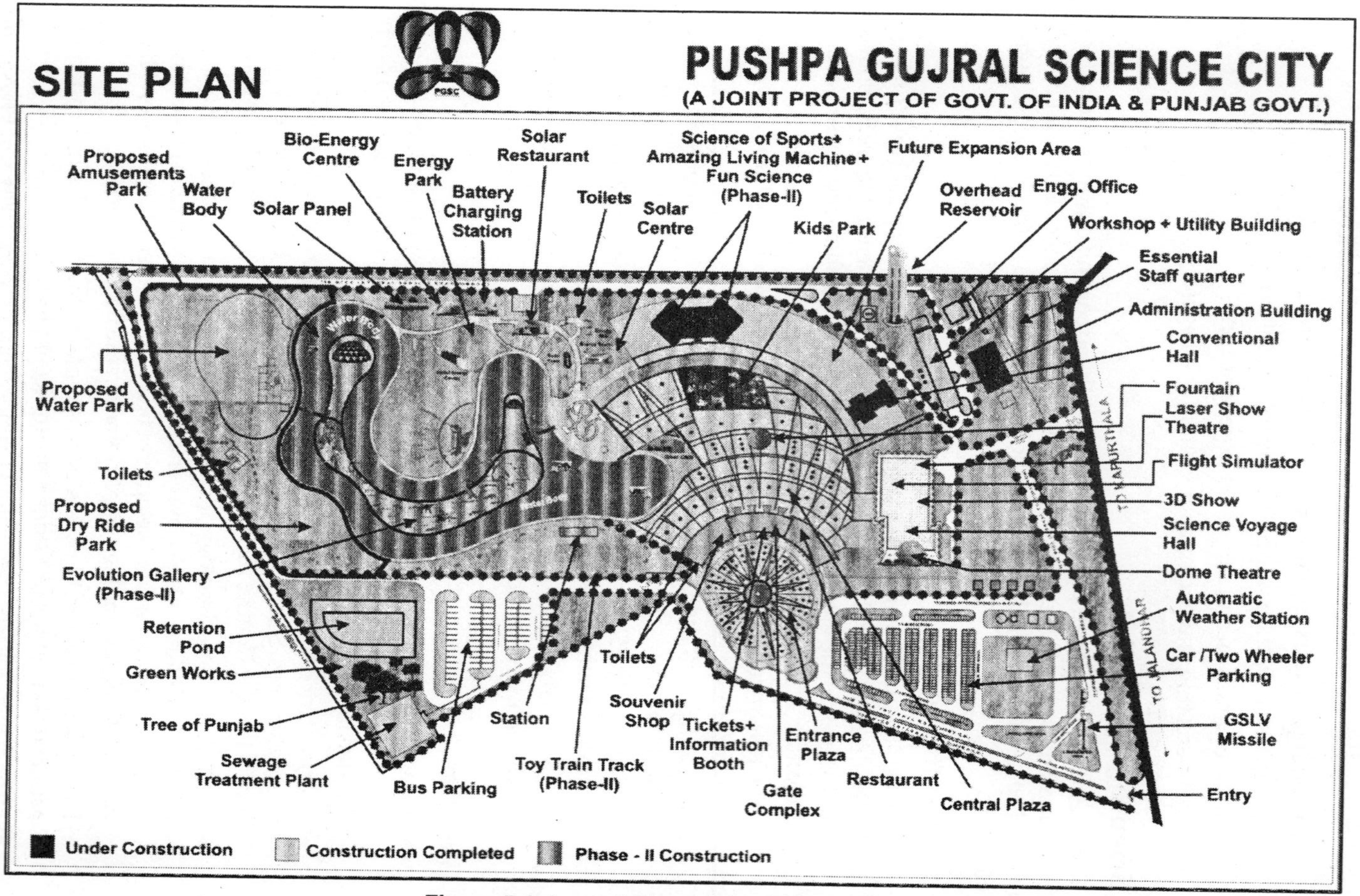

Figure 5.1: Layout Plan of the Science City

Figure 5.2: A View of the Aviation and Space Gallery and Dome Theatre Building at Pushpa Gujral Science City, Kapurthala

- Cyber Space provides hands-on experience on internet, GIS, educational software. Visitors are also exposed to exhibits on Virtual Reality.
- Science of Sports: The Gallery displays scientific principles of different sports and the physical and nutritional requirements for improved performance.
- Emerging Technologies to make the visitors aware about the latest technologies, a new gallery on Emerging Technologies has been planned. This gallery will show how nano-technology, biotechnology, optical computing and levitation etc are going to change the science scenario in the coming years.

5.3 Special Pavilions

- Railway Pavilion the Ministry of Railways would be providing an Engine Simulator, Technologies used in the Railway System and dioramas on unique hill railways. (Nilgiri Railways, Kalka-Shimla Railways and Darjeeling Railways.)
- Defense Pavilion displays a fighter Aircraft, a tank, an anti aircraft gun, a helicopter etc.
- Current Weather Monitoring Station collects data on vital weather parameters and uploads it to the Satellite. The data is processed at a central station.

5.4 An Artificial Lake (Water Body)

Approximately 1 kilometer long provides boating facilities.

6. Impact of the Science City on Sustainable Development

Various studies (Garnett, 2001) on evaluation of the impact of Science Centres and Museums the world over have established that these institutions can become a powerful engine of change in the region, in which they operate. One of the important issues under discussion at national and international forums is to sensitize the society about the challenges related to sustainable development.

These challenges include as to how we are going to feed an ever increasing world population, and also manage our land and food and energy production with increasing demands for conservation and alternative use of natural resources.

The United Nations Division for Sustainable Development (2007) lists out a number of areas coming with in the scope of Sustainable Development . However, at PGSC, the areas selected for public awareness include energy, health, biodiversity, biotechnology, rain harvesting, waste management, agriculture and environmental challenges. With the edutainment approach followed in the Science City, the public is being sensitized about the delicate balance that exists between the present day human needs and life style and preserving natural resources and echo systems. The Science City is effectively disseminating information on ways and means to meet the needs of the present without compromising the ability of the coming generations to sustain themselves.

The evaluation of the impact of the Science City indicates that it is playing a significant role in the economic development of the region (Mukhopadhyay, 2005). The contribution comes in terms of direct economic activity relating to employment generation, supply of goods and services to the Science City by various suppliers, establishment of in-campus knowledge-based commercial activities and spending by visiting public on accommodation, food and travel etc (Skane, 2004).

PGSC is focusing its attention to act as resource for

- Entrepreneurship development;
- Vocational guidance and training in the emerging areas of science and technology;
- Setting up an inventor's forum to encourage innovation and facilitating interaction between the scientists and members of the public;
- Bringing about a feeling of national pride among the visitors about the nationally/internationally renowned scientists of the country;
- Encouraging people to live with nature and making them aware of problems relating to energy and environment conservation;
- Bringing a spirit of adventure through sports and generating a pride in the local community for having a unique centre amongst them.

The healthy development of a child coupled with his/her good performance in school/college is always desired by the parents. Studies (Sharp, 2002) have shown that students in activity-based programmes exhibit increased creativity, positive attitudes towards science, perception, logic development and communication skills, leading to greater satisfaction among the parents. Also, the level of understanding of science and technology, which is now considered essential to function as citizens in a modern industrial society is positively related to the civic scientific literacy, as the Science City provides rich, social environment of the kind that support learning.

The Science City is a truly an Edutainment Centre, catering to the knowledge based entertainment needs of the community. It is attracting large crowds, specially on weekends and offers opportunities to spread the intended message on sustainable development. Figure 5.3 shows a view of the Fun Science Gallery of the Science City.

Figure 5.3: A View of the Exhibit Gallery

The following are special focus areas in the Science City for creating public awareness on Sustainable Development.

6.1 Program on Global Warming

An area of world wide interest and discussion concerns global warming or climate change which refers to rise of global temperatures caused by increased amounts of green-house gases in the atmosphere. PGSC is committed to providing individuals the opportunity to reduce the threat of climate change and is playing a pivotal role in spreading awareness about measures that can contribute to combat global warming. PGSC has started a massive campaign in this direction. Demonstrations and special programs have been developed to educate the visiting public in popularizing the renewable energy sources, conservation of water by use of sprinklers or drip-irrigation devices for watering the lawns, recycling of waste, popularizing afforestation, saving energy and reducing air pollution etc.

Special films on the effects of climate change on the melting of the glaciers in the Himalayas, rising of the sea level in the south east India and changing ecology of some parts of India are regularly screened.

6.2 Energy Education and Awareness Park

A State level Renewable Energy Education and Awareness Park has been set up at the PGSC. The main objective of setting up this park is to create awareness amongst people in general and the children in particular about the use and benefits of renewable energy technologies and to spread the message about energy conservation and efficient utilization of energy resources. India is blessed with an abundance of sunlight, water resources and biomass. The importance of increasing use of renewable energy sources as a sustainable energy base is being effectively demonstrated by the PGSC to the visitors.

The Energy Park shown in Figure 5.4 has specialized activity areas such as:

Solar Power Centre

20 KW capacity solar power centre, comprises 260 PV modules. The electricity generated is used to run Solar Fountain, Solar Toy Train and operation of Indoor exhibits in the Energy Park.

Solar Exhibit Centre

Houses exhibits operating on solar energy which include Solar Lantern, Solar Cooker, Solar Weighing Machine, Solar Still, Solar Dryer, Solar Water Heating System, Solar Dish Cooker, Solar Refrigerator and Solar Music System.

Figure 5.4: View of Energy Park

Geothermal Energy

A model of the geothermal energy demonstrates the use of heat from within the earth to generate electricity.

Wind Energy

Wind mills installed at the Science City generate electricity which is used for various applications like lifting of water.

Hydel Power Centre

This centre has exhibits to demonstrate the use of kinetic energy of moving water to produce electricity. The working models of the commonly used turbines show the generation of electricity from the moving water.

Biogas Plant

Models have been developed to explain the generation of gas from the bio-waste.

Nuclear Energy Centre

This gallery shows models of different nuclear reactors alongwith information on the need of nuclear energy and its development in India.

The visitors to the Science City are sensitized to the concept that the energy conservation is a conscious attitude, which needs to be developed, implemented and publicized wherein each one of us can effectively contribute.

6.3 Health and Bio-Technology

The feasibility study carried out in 2003 by the PGSC for identifying the galleries to be set-up in the Science City indicated that the prospective visitors would like to learn about the human body–the amazing living machine as a matter of first priory. PGSC has set-up the health gallery which not only displays the anatomical and physiological aspects of the human body, but also demonstrates the causes and remedies for the rapidly developing life style diseases. Figure 5.5 shows a view of the Health Gallery. The gallery also includes sections on bio-technology and HIV/AIDS.

In order to spread awareness about HIV/AIDS among the vulnerable sections of the society *i.e.* the adolescents studying in the colleges, Red Ribbon Clubs have been set-up to train peer groups. A specially designed Bus along with two counselors–one male and one female–goes to the colleges regularly. The films on HIV/AIDS are screened on a LCD monitor fitted in the Bus. The project has received tremendous response from the students and applause from the health authorities. Figure 5.6 shows the Red Ribbon Express Bus and its interiors fitted with LCD monitor.

6.4 Waste Management

The underground sewage collection and distribution system incorporates discharge to a Sewage Treatment Plant, which forms a part of 'Kidneys of the Site'. The other facilities in the section include Retention Pond, Composting, Recycling, Sewage Treatment and Sludge Tank as a behind-the-scenes experience for specially guided groups to show how Science City responsibly and innovatively deal with its waste stream.

Figure 5.5: View of Health Gallery

Retention Pond is configured to function as a soakage pit to replenish the subsurface ground water aquifer. There is a proposal to make the following provisions for waste management:

Composting bins are being constructed to take all vegetable waste and turn it into enriched soil that can be reused in the gardens.

Recycle bins are located to recycle glass, paper, metals and plastics into a recycling waste reuse stream.

PGSC is reusing and disposing the "sludge"–a by-product of the sewage treatment process. The sludge is suitable for application in compositing, pressing and drying for reuse as fertilizer on site. Figure 5.7 shows the Sewage Treatment Plant.

6.5 Environmental Considerations

Special Care has been taken to prevent and control air pollution in the Science City campus. Exhaust emissions from two sources, vehicles carrying visitors to and from the site and the stand-by/emergency power diesel gen-sets are the principal sources of air pollutants.

The parking facility has been located at the immediate arrival site of the visitors. The location combined with the parking facility design minimizes travel distance

Figure 5.6: Red Ribbon Express Bus (HIV/AIDS Awareness Programme)

Figure 5.7: Sewage Treatment Plant

resulting in reduced vehicle emissions. In addition, the parking facility design incorporates extensive landscape and shaded tree plantation to assist in point source absorption of vehicle emissions.

The stand-by/emergency power gen-sets, which have been placed in a remote, unseen-by-visitors location, are provided with flexible engine exhaust piping that has a noise suppression muffler system and exhaust filters.

6.6 Water Management

The unique aspect of the PGSC is that it is not just a science-centre-in-a-box. It is spread over 72 acres of land that enables the programmed environmental and water management education opportunities to the visitors.

The project site is designed to be fully self sufficient with potable water supply via on-site tube wells, complete on site harvesting, retention and ground water recharge of storm water and treated waste water due to planned recycling of water from the sewage treatment plant, to maximizes the potential to recycle available water resources.

The visitors are demonstrated the use of sprinklers and drip irrigation methods to save water. The visitors are also exposed to the management of water in a typical urban water supply system.

6.7 Bridging the Digital Divide

The 'Digital Divide' has started making distinct disparities between the developed countries and the developing world. Many people in the developing countries do not have access to computers for numerous reasons as diverse communities, people with low incomes and the less educated are the groups often lacking access to information resources. Also, disparities exist between different sections of the society and different regions. As much as the new technology can equalize opportunities for students of all backgrounds, the absence of these basic resources–computers, software and internet service–has the potential of widening gaps between populations.

To develop new products and stay competitive, high technology industry demands computer skilled workforce. Education is always helpful when it comes to getting good jobs. The modern economy is increasingly based on information technology and getting a job may depend upon how technically skilled you are. Jobs requiring no computer skills will become fewer and fewer in the times to come.

The Science City offers immense opportunities for bridging the digital divide by providing facilities to the local population by offering basic computer courses and equipping them with the adequate proficiency in computer literacy. Such an experiment has been tried at the Science City where the people of all ages (from 10 years to 70 years) have shown tremendous interest in undergoing basic computer courses. Figure 5.8 shows a senior citizen undergoing computer training at the Science City.

Figure 5.8: A senior citizen undergoing training in the Computer Centre

6.8 Encouraging Use of Internet

Science City is spreading awareness about the usage of Internet, particularly for those citizens who have already crossed the formal education age. As we know, the Internet is changing how we communicate, shop, educate ourselves, get our news and elect leaders. Even jobs like fashion designers, musicians, construction workers and auto-mechanics require computer knowledge and internet consultation.

The Science City is playing an important role in encouraging young people to become technologically literate and be able to browse professional and community based sites. Open access to internet is provided in the Science City to motivate the general public about the immense potential the internet offers in today's world.

6.9 Strengthening Knowledge-Based Society

We are living in an era of the information revolution and the information age which is fast resulting in an on-going process of economic, social and political globalization. One critical issue that emerges from all of these processes is the central role of knowledge, education and learning for the success of the global information society and global information economy. Knowledge is becoming an increasingly important factor of production even more important, some analysts would argue, than land, labor and capital. Almost all developing countries are adjusting their priorities to strengthen the basis for knowledge society and knowledge management. The Science City is playing an important role in strengthening knowledge based society through a multitude of programs (Garnett, 2001).

6.10 Global Perspective

A principle challenge to education today is to shift from a localized factory-model focus to one that prepares citizens to function in an increasingly interconnected and rapidly changing globalized world (Smith, 2002). The Science City is playing an effective role in addressing global problems such as environmental issues and concerns from population growth to climate change and from HIV/AIDS to Biotechnology. All these demand the cultivation of interdisciplinary and global perspectives to counter and share global human problems.

7. Celebration of Important Science and Technology Related Events

The objective of the programme is to create awareness about various facets of science, technology and environment amongst students, youth, teachers, farmers, professionals and the general public etc. In these programmes, the societal issues such as environmental protection and energy management are discussed with the active involvement of the community (Walter, 2001). The Important days which are celebrated in the Science City are:

World Wetland Day (February 2)

On this day, in 1971, the Ramsar Convention on Wetlands of International Importance was signed. Wetlands are a very important part of our biodiversity and it is essential to see that they are well protected.

National Science Day (February 28)

National Science Day is celebrated to commemorate the discovery of the Raman Effect for which Dr. C.V. Raman, an Indian Scientist, was honoured with the Nobel Prize.

World Forestry Day (March 21)

Activities such as the planting of trees and highlighting the urgency to increase the green cover is portrayed with the help of audio-visual programs.

World Water Day (March 22)

The day is celebrated to highlight that the drinking water sources are fast depleting and there is a need to conserve the same.

World Health Day (April 7)

The World Health Organization (WHO) was constituted on this day in 1948. Every year, the WHO puts forward a theme-specific health related issue for bringing about public awareness. The day is celebrated in the Science City with great enthusiasm.

Earth Day April 22

The day is celebrated to draw the attention of the world to the problems being caused to the earth due to modernization.

Anti Tobacco Day (May 31)

The world is now aware of the problems faced by not only the smokers but also the people who inhale the smoke. Bringing public awareness and launching anti-smoking campaign among the visitors is strongly advocated in the Science City through film shows and lectures.

World Environment Day (June 5)

This day is celebrated as a consequence of the Stockholm Conference on Human Environment held in 1972. Concerns for the increasing environmental problems are communicated to the visitors on this day.

World Population Day (July 11)

Awareness about the growing problems with the increasing Population needs to be given special attention, especially in India with its an ever-increasing population problem.

World Ozone Day (September 16)

The United Nations declared this day as the International Day for the Preservation of the Ozone Layer. It is the day the Montreal Protocol was signed. Lectures on the importance of the day are organized to bring about public awareness on this vital issue.

Green Consumer Day (September 28)

The problems of consumerism and its impact on the environment is an area of major concern in today's world. Awareness building programs on the importance of recycling-reusing-reducing the consumer goods waste are held on this day.

World Habitat Day (October 3)

The earth is the habitat of not only human beings but also all living creatures. Increasing human activities is threatening the habitat of other living things. The visitors are sensitized to the issue of preservation of bio-diversity.

World Wildlife Week (October 1-7)

The week is celebrated by building awareness on the importance of preservation of our wildlife.

World Animal Welfare Day (October 4)

The day is celebrated to highlight the importance of welfare of animals.

International Day for Natural Disaster Reduction (October 13)

Due to several reasons, there has been an increase in the number of natural disasters. Public is educated on the disaster mitigation measure.

Children's Day (November 14)

Children can work together for a better tomorrow by improving the environment around them. The day is celebrated by arranging special programs for Children in Science City.

On these days, the following items are organized to bring about public awareness of the burning issues:

- A popular lecture on the theme of the DAY
- Putting up Information panels along with photographic/illustrative enlargements
- Enacting a play highlighting the issue.
- Organizing a quiz competition.
- Organizing debate on the particular issue.
- Organizing Poster competition.
- Having a photograph competition.

8. Conclusion

All nations face a broad array of challenges that require the application of scientific knowledge and technology for sustainable development. Such challenges include stimulating economic growth, mitigating environmental problems, safely adopting beneficial new technologies, and quickly responding to sudden outbreaks of new diseases. No nation can now afford to be without access to a credible, independent science and technology (S&T) capacity that would help it to develop informed policies and take effective action in these areas. Science City is making efforts to play an effective role to build a strong and confident relationship with the community so as to jointly work towards ensuring Sustainable Development.

References

Garnett, R., 2001. The Impact of Science centres/Museums on their Surrounding Communities: Summary Report.

http://www.astc.org/resource/case/Impact_Study02.pdf (accessed on 4 Feb. 2008)

Mukhopadhyay, I.K., 2005. The Science Centre Movement in India–A Conspectus. 4th Science Centre World Congress.

http://www.scielo.br/pdf/hcsm/v12s0/13.pdf (accessed on 4 Feb. 2008)

Sinha, A., 2005. National Council for Science and Technology Communication (NCSTC), Department of Science and Technology, Govt. of India.

http://dst.gov.in/scientific-programme/s-t_ncstc.htm (accessed 30 Dec. 2005)

Skane, W., 2004. Strong Science and Technology Capacity a Necessity for Every Nation. U.S. National Academy of Sciences, Washington, D.C.

http://www.interacademycouncil.net/?id=10011 (accessed 4 Feb. 2008)

Smith, M.K., 2002. 'Globalization and the Incorporation of education' the encyclopedia of informal education.

www.infed.org/bibilio/globalization.htm (accessed 30 May. 2005)

Sharp, Lindsay (2002): Science Museum Visitor Brochure published by NMSI, UK

http://www.nmsi.ac.uk

United Nations Division for sustainable Development. Documents: Sustainable Development Issues

(http://www.un.org/esa/sustdev/documents/docs_sdissues.htm) Retrieved: 2007-05-12

Walter, W., 2001. Many Roles to Play: The Science Centre as Community Powerhouse. Association of Science–Technology Centres.

http://www.astc.org/pubs/dimensions/2001/jan-feb/community_powerhouse.htm (accessed 28 Dec. 2005)

Chapter 6

Outreach Policy of Science Centres

Iskandar
Pusat Peragaan IPTEK, Science–Technology Centre,
Jakarta 13560, Indonesia
E-mail: uan_is@yahoo.co.id, isk@ppiptek.or.id

ABSTRACT

The capability of society in the development and application of science and technology is an important key to develop a country. As a fast country with huge population spread in thousand of islands in 33 provinces, Indonesian government needs to apply the principle of equity in implementing the development program all over Indonesia. Indonesia's people have the same opportunity in the acquisition and sharing the benefits of the development results. So the development could not be applied to one region only but throughout Indonesia for the prosperity of the whole nation. As of today Indonesia has several Science Centres spread out in several Provinces namely, Jakarta, Bandung (West Java), Palembang (South Sumatera), Sidoarjo (East Java) and Yogyakarta. It means that most of the Provinces do not yet have a Science Centre. Based on the reasons above, more science centres are needed. The policy of the Indonesian government is to develop science centres at least one science centre in the capital city of each province refers to State Ministry of Research and Technology Decision number 75/M/Kp/IX/2001. The Ministry of Research and Technology of Republic of Indonesia as the coordinator of program assigns Science and Technology Centre Indonesia (PPIPTEK Jakarta) to do pioneering work for new science centres in other provinces in Indonesia. One of the most important tools considered is the outreach program (traveling exhibition). PPIPTEK Jakarta uses outreach program to attract, to motivate and to get support from all parties of the society in those provinces for developing science centres. This paper would present the role of Science and Technology Centre Indonesia (PPIPTEK Jakarta) to support the Government of Republic of Indonesia in the development of new Science Centres in other provinces.

Keywords: *Developing country, Equity base, National development, Outreach program, Traveling exhibition, Province science centre.*

1. Introduction

Indonesia is the biggest of country Archipelagic in the world, having 17,504 islands which unfolding as long as 3,500 sea miles. Wide of Indonesia continent now are 1,904,556 km^2. Indonesia has 33 Provinces. 238.5 million people live in 349 Sub-Provinces and 91 Towns.

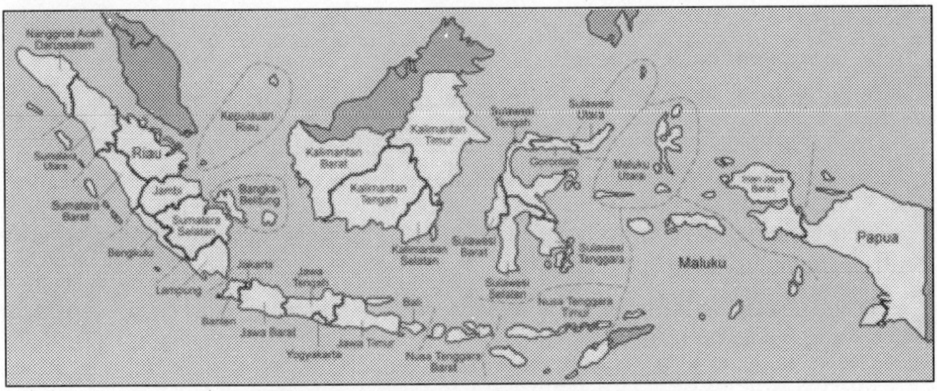

Figure 6.1: Map of Republic of Indonesia

The policy of government in science and technology is that every people in all regions have to get same opportunity in knowledge mastering of science and technology. But since economic crisis knock over Indonesia in 1998, the policy becomes executed difficult with system centrally by the government. Hereinafter, local government is very expected to be able taking a role in popularizing of science and technology for the local society refers to State Ministry of Research and Technology Decision number 75/M/Kp/IX/2001

2. Main Body/Materials

PPIPTEK Jakarta is the first Science Centre in Indonesia, founded in 1991 by Ministry of Research and Technology Republic of Indonesia, Mr. Prof. BJ Habibie. As the first Science Centre, PPIPTEK Jakarta has visited by many people from various areas in Indonesia. More than 300,000 visitors every year visit PPIPTEK Jakarta.

In addition to serving Jakarta's society in popularizing science and technology, PPIPTEK Jakarta and Ministry Office of Research and Technology Republic of Indonesia also have a role together to motivate other local government in Indonesia to build a new Science Centre. The Government of Republic of Indonesia expects that every province have a science centre to serve the local society.

To make local government enthusiasm to build a new science centre, hence the local government has to be given image about what the Science Centre is. Outreach program approach is the answer for it.

Outreach program is an approach to bring a packet of exhibition consisting of 70 equipments traveling exhibits to other province in Indonesia. Then, the traveling exhibits are arranged into a hall about 1,000 m^2 like similar to showing a mini science centre. Not only equipments of the traveling exhibits show up, but also science workshop and science demonstration activities are showed there.

For execution during 5 days exhibition are usually site attended more than 10,000 visitor, most of them are students group and his family.

3. How Does Outreach Program have a Role?

There are several elements that determine the success of science centre development in a province, those are:

1. Support of local society.
2. There is a group of people that aware to science advancement.
3. Support of local government and stake holders.
4. Support of company with it is Company Social Responsibility programs.

These elements must to have same vision and desire for developing science centre in their province.

It will be appear if they see and feel directly the advantages of science centre for science and education advancement.

So, it is time to PPIPTEK Jakarta for playing a role to introduce what Science Centre is. Through outreach program, PPIPTEK brings always at least 70 hands-on exhibits, science workshop, science demonstration, etc. With invite all of society levels, local government, stake holders, and private company, hope it can arouses and generates enthusiasm to build Science Centre in their province.

The flowchart is a new concept of idea that could be developed by the government to implement the policy describes the process of pioneering province science centre development. The first idea may come from central government in this case is State Ministry of Research and Technology, PPIPTEK, or science observers. Then it will be socialized to local government, stake holders, private company, and local society. The success of outreach program can be seen with formed a group off local society who will consistent continually to struggle Science Centre development.

4. Results and Discussion

PPIPTEK Jakarta and Ministry Office of Research and Technology have been target planning to develop 9 Science Centres in the Capital Cities of 9 Provinces till year 2010. Except Balikpapan (East Kalimantan), each Province have got twice opportunity outreach program in different places in the region of.

Nowadays Indonesia has four Science Centres exclude PPIPTEK Jakarta, namely: Sundial Bandung (West Java), Graha Teknologi Palembang (South Sumatera), Mpu

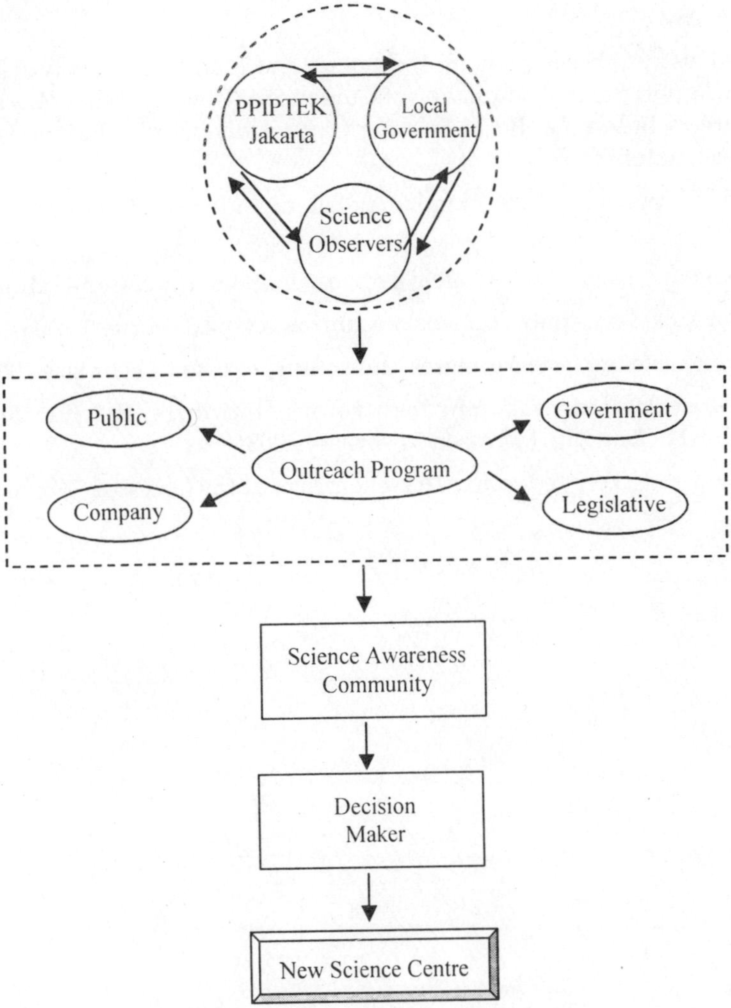

Figure 6.2: Flowchart of Outreach Program

Tantular Science Centre Sidoarjo (East Java) and Taman Pintar Yogyakarta (Central Java).

Two Provinces are conducting visibility studies, they are Medan (North Sumatera) and Balikpapan (East Kalimantan). While, three others are in process final development, they are Pangkal Pinang (Bangka–Belitung), Pekan Baru (Riau) and Kendari (South East Sulawesi).

5. Conclusion

Outreach program is one of the effective ways to approach all parties of society, to arouse science awareness, to popularize science and technology and to develop new Science Centres.

References

Team of Ministry Office Research & Technology and Team of University of Indonesia, 2006, Report Study Compilation of Guidance of Development of Local Science Centre in Indonesia, Pusat Peluang Carya University of Indonesia, Depok, JABAR, Indonesia.

http://id.wikipedia.org/wiki/Jumlah_pulau_di_Indonesia (accessed 4 February 2008)

http://id.wikipedia.org/wiki/Daftar_provinsi_Indonesia (accessed 4 February 2008)

http://www.astc.org/pubs/dimensions.htm (accessed 3 February 2008)

http://www.astc.org/sciencecentres/index.htm (accessed 3 February 2008)

http://www.californiasciencecentre.org/Exhibits/ExhibitsForRent/ExhibitsForRent.php (accessed 5 February 2008)

http://www.mdsci.org/educators/travelingscience.html (accessed 5 February 2008).

APPENDIX 1
Outreach Program

APPENDIX 2
Science Centres in Indonesia

PPIPTEK–Jakarta

Sundial–Bandung (West Java)

Graha Teknologi–Palembang (South Sumatera)

Taman Pintar–Yogyakarta (Central Java)

Chapter 7

Science, Religion and History: A Case Study

Phyllis Lam Li Wan
National Science Centre, Ministry of Science, Technology and Innovation, Kuala Lumpur, Malaysia
E-mail: phyllis@psn.gov.my

ABSTRACT

There is an urgent need to change the mindset of Malaysians especially the Muslims to maintain balance in the pursuit of knowledge-not to pursue spiritual/ revealed knowledge with the exclusion of acquired knowledge (science and technology). This is particularly critical as it is recognized that to succeed in today's knowledge intensive economy, Malaysia must be prepared to develop the skills of her citizens, promote innovation and productivity and encourage experimentation–to be actively engaged in R&D. "The Scientific Excellence in Islamic Civilization Exhibition, Kuala Lumpur 2007" which was developed and produced in collaboration with the Institute for the History of Arabic- Islamic Sciences, Johann Wolfgang Goethe University Frankfurt is one such platform which the Government of Malaysia believes would help generate greater awareness, inspiration and appreciation of the Muslim scientific discoveries and their legacy of inventiveness and scholarship to the Muslim and non-Muslim community in this country. It is hoped that through well planned and structured programmes, the Government would be successful in resolving this false perception and changing this mindset among Malaysian Muslims in order to remain on track to attain a developed country status by 2020. The paper continues to discuss further on the exhibition, the education and public programmes and– the outcome, experience arising from organizing this large scale exhibition programme.

Keywords: *Scientific excellence, Islamic civilization exhibition, Mindset of Malaysians, Road show.*

1. Introduction

"The Scientific Excellence in Islamic Civilization Exhibition, Kuala Lumpur 2007" was developed and produced in collaboration with the Institute for the History of Arabic-Islamic Sciences, Johann Wolfgang Goethe University Frankfurt. The Minister of Science, Technology and Innovation, the Honourable Dato' Seri Dr Jamaludin Jarjis was instrumental in bringing this exhibition to Malaysia and was involved in the process every step of the way. There were two main components to the exhibition programme comprising of the exhibition (at two consecutive venues ie. Kuala Lumpur Convention Centre and the National Sceince Centre) and a roadshow.

2. The Rationale

It is recognized that the timing of this exhibition was most appropriate at a time when the Muslim Ummah (group of believers in Islam) is seen to be lagging behind in the fields of science and technology in this modern world. The purpose of this exhibition would be to remind the world of the richness and openness of the Muslim scholars during the earlier centuries when Islamic scholarship was at the pinnacle of its enlightenment. The Government of Malaysia believes that exhibitions such as this one would generate greater awareness, inspiration and appreciation of the Muslim scientific discoveries and their legacy of inventiveness and scholarship to the Muslim and non-Muslim community in this country.

There is an urgent need to heighten and convey awareness among the Muslim and non-Muslim community on the pioneering spirit of the Muslims which led to breakthroughs in science which eventually led to the renaissance especially of science in the western world. To succeed in today's knowledge intensive economy, we must be prepared to develop the skills of our citizens, promote innovation and productivity and encourage experimentation. A nation filled with constant flow of new ideas and innovations will emerge as a contributor and leader in technologies. Thus, investments in knowledge-seeking activities such as science and technology including research and development are critical in ensuring a constant flow of ideas from which we can reap enormous benefits.

In Malaysia, a country with about two-thirds population who are Muslims, and as in many other Muslim countries, there seems to be a false perception that a Muslim's pursuit and search for knowledge should be totally and exclusively that of revealed knowledge.

Other forms of embodiment of acquired knowledge are met with resistance and at times deemed to be worldly, secular and in some extreme cases as being unIslamic. Hence, it will be Malaysia's greatest challenge to urgently resolve this false perception and change this mindset in order to remain on track to attain a developed country status by 2020.

3. Objectives of the Exhibition

The objectives of this exhibition in Malaysia were:

(*i*) To change the mind-set of Malaysians regarding the importance of acquiring scientific knowledge and technical know-how. There is a need to dispel the false perception of the Muslims in Malaysia that Islam is confined only to spiritual pursuits and pursuits of the revealed knowledge. Islam requires the ummah to pursue knowledge embodied in both revealed and acquired knowledge;

(*ii*) To inform and educate the public about the 'Golden Age of Islamic Civilization', whereby many great scientific and technological advances made by Muslim scholars were assimilated and became integral to the Western Civilization.

(*iii*) To inspire and motivate the main target group, that is, the younger generation to appreciate the scientific discoveries Muslims have made over the years–to understand the struggles faced by the ummah, their perseverance and determination in producing scientific and technological advances in an era where Muslims work harmoniously with people of different faiths and races worldwide.

4. "The Scientific Excellence in Islamic Civilization" Exhibition in Kuala Lumpur 2007

This exclusive exhibition was a collaborative effort between the Institute for the History of Arabic-Islamic Science, Johann Wolfgang Goethe University, Frankfurt, Germany and the Ministry of Science, Technology and Innovation (MOSTI) Malaysia.

Exhibition Concept

The exhibition was divided into five main areas of focus: Astronomy, Geography and Navigation, Medical Inventions and Discoveries (including Chemistry), Technology and Inventions (including Physics), and Mathematics, Art and Architecture.

The visitors are first greeted by a majestic city gate before being invited to step back 1000 years in time into the world of ancient Islamic civilization.

Purpose-built props serve as a backdrop and provide the setting for exhibits in the different areas of focus:

- An interactive constellation dome, sundial (with simulation of moving sun) in Astronomy
- An ancient ship–where visitors can go on deck to view the displays/ instruments in Geography and Navigation
- Facade of an ancient hospital building–to display medical instruments, alembic vessels in Medical Inventions and Discoveries
- Graphic mural wall of ancient agriculture scene and inventions with exhibits being showcased in display cases in Technology and Inventions
- Facade of a city wall depicting Muslim architecture and design in Mathematics, Art and Architecture area
- A mini-theatre where available documentaries on the contributions and achievements of Islamic Civilization are shown.

Exhibits

The exhibition featured a total of 148 instruments and working models which were on loan from the Institute for the History of Arabic-Islamic Science, Johann Wolfgang Goethe University, Frankfurt, Germany. With the advice from Prof Fuat Sezgin, these instruments and models were displayed in the five main areas of focus.

In addition to the 148 exhibits, the National Science Centre fabricated a number of interactive exhibits such as an interactive timeline of the scientific and technological discoveries and inventions covering the pre-Islamic era, Islamic civilization and post-Islamic period, a sundial which simulated the workings of an actual one and an interactive constellation dome where visitors were able to locate the stars/ constellations discovered by Muslim scientists. To further enhance and deepen visitors' understanding of the exhibition, a number of multimedia programmes based on significant discoveries and inventions in the various fields of astronomy, geography and navigation, medicine, mathematics, art and architecture, technology and inventions, an interactive digitized version of Al-Jazari book of inventions, and video presentations on the contributions and achievements in science and technology during the Islamic civilization were also provided.

Exhibition

It took five days and thirty people working round-the-clock to set-up the exhibition at the Kuala Lumpur Convention Centre (KLCC), a mammoth task indeed. Nevertheless, their spirits were high and the air was buzzing with excitement and anticipation. The whole process was likened to putting a puzzle together. Finally, everything was in place and ready for the opening.

It was indeed a grand opening of the exhibition at KLCC which was officiated by the Prime Minister of Malaysia, the Right Honourable Dato' Seri Abdullah Ahmad Badawi. Throughout the one month duration of the exhibition at KLCC, the exhibition received visits from a wide spectrum of visitors–from royalty to school children. Dignitaries and their entourage who visited the exhibition included the King of Malaysia, Sultans, Governors, Chief Ministers, Cabinet Ministers, and political leaders.

Complementary Activities

Activities were planned to meet the needs and interest of the whole spectrum of visitors to the exhibition. The scientific findings of Islamic scholars presented in a fun, interactive way through science demonstrations and many hands-on educational activities drew the attention and participation of school children. The educational activities included: how a camera obscura works, optics (Ibn al-Haitham), raising water–pressure and moments (al-Jazari), chemistry–acids and alkali, taxonomy–classification of plants (Ibn-al-Baytar), production of perfumes from distillation techniques (Ibn Hayyan, al-Kindi). Discussions followed the activities to relate experiments to the discoveries and inventions of the Islamic scholars. Tired visitors could also gain respite at the Islamic garden and browse through the many books and reading materials on Islamic scholars and their contributions.

A special booth outside the main exhibition area was set-up to showcase Malaysia's efforts in aerospace. The highlight of this display was the Malaysia's astronaut programme which generated a great deal of interest and drew attention of the visitors.

Daily lucky draws and weekend Nasyid performances also proved to be popular with the visitors and served to entertain the crowd waiting to view the exhibition.

Facilitators

To further enrich visitors' experience of the exhibition, a total of 20 University graduates were selected and engaged as exhibition facilitators to guide visitors through the exhibition. These facilitators received comprehensive training on the historical and scientific aspects of the instruments and devices on display including the storyline and flow of the exhibition from the institute's curator, Mr Lutz Kotthoff and the National Science Centre's officers.

Visitorship

The exhibition received very good response from the public and visitor numbers are as follows:

Venue	*Duration*	*Visitor Nos.*
Kuala Lumpur Convention Centre	8/1–14/2/2007	225,841
National Science Centre	23/2–31/3/2007	30,811

5. On-Wheels Exhibition

While the exhibition was running at KLCC, MOSTI also embarked on programme called "On-Wheels Exhibition" to:

(i) Showcase a selected number of representative exhibits from the original 148 instruments; and

(*ii*) Display the interactive multimedia and audio visual programmes.

The prime objective to have this mobile exhibition-on-wheels was to reach out to the rural community particularly the Muslims who were unable to visit the exhibition in Kuala Lumpur carrying the same message to visitors to the exhibition.

These exhibits were fitted into two modified buses–one multimedia interactive bus equipped with 15 multimedia stations installed with interactive programmes on the scientific contributions and inventions–astrolabe, maps, compass, water-raising machines, architecture and the other an exhibition bus with 27 exhibition stations. It was planned that the buses travel together in a convoy fashion, park at selected mosque premises in the rural community throughout the country and run for duration of two days per venue.

The road show ran on a very tight schedule with an allowance of one day for travel and set-up time at each venue, meaning that there was practically no room for hiccups. However, with the close cooperation from the state governments which assisted with the local logisitic arrangements. This "On-Wheels Exhibition" was a huge success and received a total of 14,929 visitors covering 11 states in Peninsula Malaysia.

6. The Next Step

The exhibition has been developed further as a thematic exhibition and plans are underway to exhibit it in various venues in Malaysia.

There are plans to establish an Islam Hadhari Science and Technology Centre in Kuala Lumpur.

7. Conclusion

It was indeed a very successful exhibition which touched the minds and hearts of Malaysians. Our Prime Minister was most impressed with the reception and had expressed his wish for Malaysia to continue further collaboration with the eminent Prof. Dr. Fuat Sezgin to have a permanent exhibition through the establishment of an Islam Hadhari Science and Technology Centre in Kuala Lumpur.

It is recognized that with continued efforts to bring this exhibition and the underlying message to other Islamic countries, we will be able to change the mind-set of Muslims that pursuits of acquired knowledge like science and technology is important, is not secular, unIslamic, conversely, it is fundamental to the development of their social well-being and critical to the nation's progress and active participation in the new economy.

References

Abbot, A., 2004. Islamic Science: Rebuilding the past. News feature, Nature, Vol.432, pp.794-796.

Morrison, A., 2005. Ahead of their time. Time/Paris, Sunday Nov.13.

http://www.time.com/time/magazine/article/0,9171,1129488,00.html

Sezgin, F., 2004. Science and Technology in Islam. Catalogue of the Exhibition of the Institute for the History of Arabic-Islamic Science (at the Johann Wolfgang Goethe University, Frankfurt, Germany), Frankfurt Book Fair 2004.

Chapter 8

Rajiv Gandhi Science Centre: Contributing to the Development of Mauritius

Aman Kumar Maulloo

Rajiv Gandhi Science Centre, Old Moka Road, Bell Village, Mauritius

E-mail: rgsctf@intnet.mu

ABSTRACT

The Rajiv Gandhi Science Centre (RGSC), a parastatal body under the aegis of the Ministry of Education and Human Resources[1], is the unique Science Centre in Mauritius. It has as one of its main objectives to inculcate scientific awareness in current and relevant fields and arouse curiosity among the public in general. Since its inauguration in November 2004, the Rajiv Gandhi Science Centre has been striving in this direction to realise its vision of becoming a Centre of Excellence in the communication and promotion of Science and Technology. This paper starts with an overview of the growing importance of science in the socio-economic development of Mauritius. The status of science in our educational system and the bold decision by the government to boost the teaching of science in our schools is highlighted. The introduction is followed by the actions taken by the Rajiv Gandhi Science Centre to popularise science among the public at large. Besides its five interactive galleries, the types of activities have ranged from educational talks and workshops to competitions and exhibitions at national level. In view of the fast evolution of Science and Technology in the world, there is a need for Science Centres to adapt dynamically in order to assist the country achieve their set objectives. The Rajiv Gandhi Science Centre is therefore implementing a series of measures to promote Science and Technology at various

1 Since September 2008, the Rajiv Gandhi Science Centre falls under the aegis of the new Ministry of Industry, Science and Research. Further, the Ministry of Education and Human Resources is now known as the Ministry of Education, Culture and Human Resources.

levels among the Mauritian population to be in-line with the Government policy. Some measures that are presented include the development of the Mauritius Science Portal, contribution to the Knowledge Channel of the national digital television, outreach programmes, and tailor-made programmes for specific sectors of the Mauritian population.

Keywords: Rajiv Gandhi Science Centre, Science education, Science park, Exhibition galleries, Popularisation of science, Knowledge Channel, Knowledge-based economy.

1. Introduction

Mauritius is an island of about 1800 Square kilometers situated in the Indian Ocean approximately 2000 km off the South East coast of Africa. It was discovered initially by Arab sailors and then was successively occupied by several European settlers including the Portuguese, the French and lastly by the British since 1810. The island obtained its independence on the 12th March 1968 and was proclaimed a republic in 1992 (Selvon, 2001).

The present population of 1.2 million forms a mosaic of different races, cultures and religions as Mauritians are descendants of immigrants from the India, Africa, Europe and China respectively. The cultural diversity and racial harmony of the island is one of the most valuable inherited assets of Mauritius. Most Mauritians are multilingual, being fluent in Creole, French and English. English is the official language. Bhojpuri, Hindi, Urdu, Tamil, Marathi, Telugu and Mandarin are also spoken.

Mauritius is a democracy modelled on the British system of parliamentary democracy, which guarantees the separation of legislative, executive and judicial powers. The President is the Head of State while the Prime Minister has full executive powers and is the Head of Government. Sixty-two members of the National Assembly are elected every five years by universal adult suffrage. Democracy is well entrenched in Mauritius and all major political parties are represented in Parliament.

A series of economic reforms have been implemented to gear Mauritius towards growth and financial sustainability. Mauritius has evolved from a mono-crop industry to a multi-pillar economy with tourism, textile, and the financial services as the main supports. The government is working towards the democratization of the economy to meet a multitude of objectives including land reform, improvement of institutional framework and promotion of entrepreneurship. Measures are undertaken to establish ideal business environments to attract foreign investors (Government Programme, 2005-2010). Since we are encouraging foreign investors, there is a need to develop a critical mass in various fields of Science and Technology to reduce unemployment rate while minimizing the existing job mismatch.

An important amount of the annual budget is absorbed in the education sector. While education is completely free up to secondary level, free transport is also provided to students and senior citizens. The main aim is to make education accessible to one and all and at the same time to provide a World Class Quality Education (Government Programme, 2005-2010).

This paper starts with an overview of Science Education in Mauritius. The status of science subjects in our schools is highlighted along with the measures that are undertaken to boost the teaching and learning of science. This is followed by the role of the Rajiv Gandhi Science Centre to promote Science to the public at large. Finally, the various strategies employed by the centre both independently and in collaboratively with other local institutions to support Mauritius to become a Knowledge Hub in the region are covered.

2. Status of Science in Mauritius

With a view to assess the status of Science and Technology in Mauritius in general, and more specifically in the educational sector, the Mauritius Research Council (MRC)[2], a parastatal body falling under the aegis of the Prime Minister's Office, in its capacity as the advisor to Government on matters pertaining to science and technology, undertook a number of studies. The findings of these studie's were alarming as it showed clearly that not only science was becoming less attractive in schools but public understanding of science was on the decline (MRC, 2004; Suddhoo, 2007).

Figures 8.1 and 8.2 indicate that while more boys are enrolling for science subjects at O-Level, girls are less likely to opt for Physics and the boys' tendency to do Biology is on the decline. It may be noted that at most 20 per cent of the total students choose science at O-Level. On the contrary the performance of girls surpasses that of boys, particularly in Physics, showing that girls do possess a strong affinity for science.

The trend at A-Level is more worrying as indicated in Figure 8.3. This chart depicts the preference for boys and girls combined to opt for Mathematics but it also

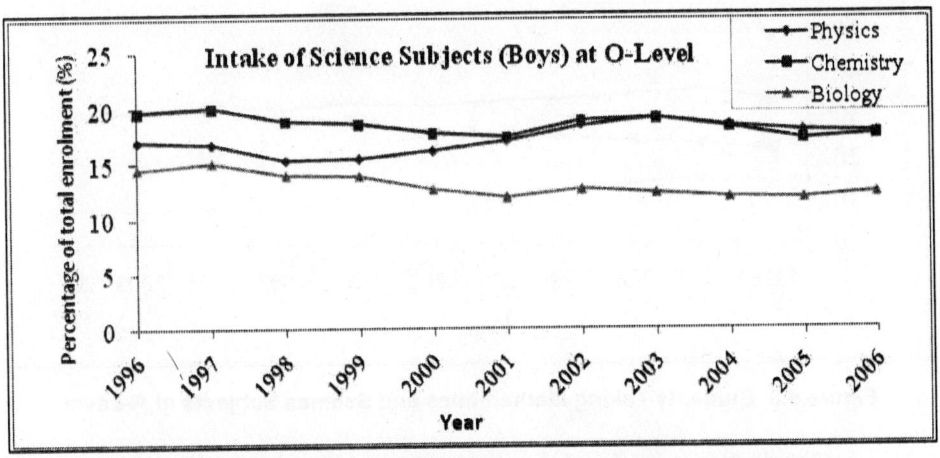

Figure 8.1: Percentage of Boys Taking Science Subjects at O-Level

---

2 Since September 2008, the Mauritius Research Council operates under the Ministry of Industry, Science and Research.

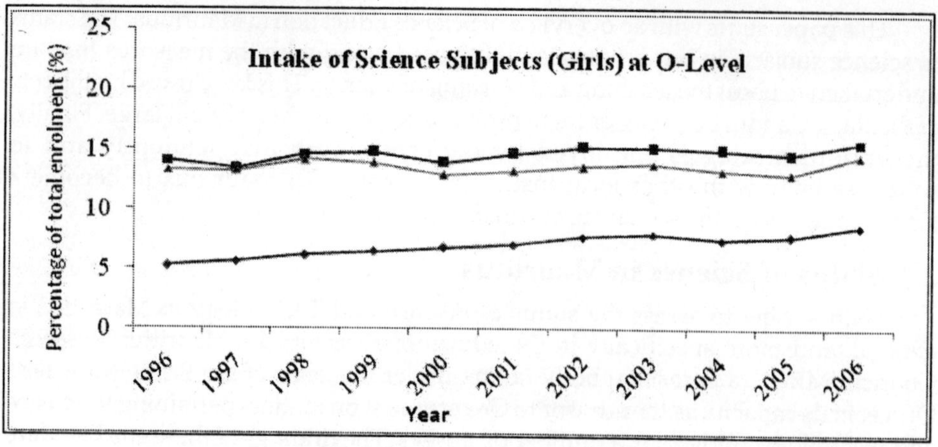

Figure 8.2: Percentage of Girls Taking Science Subjects at O-Level

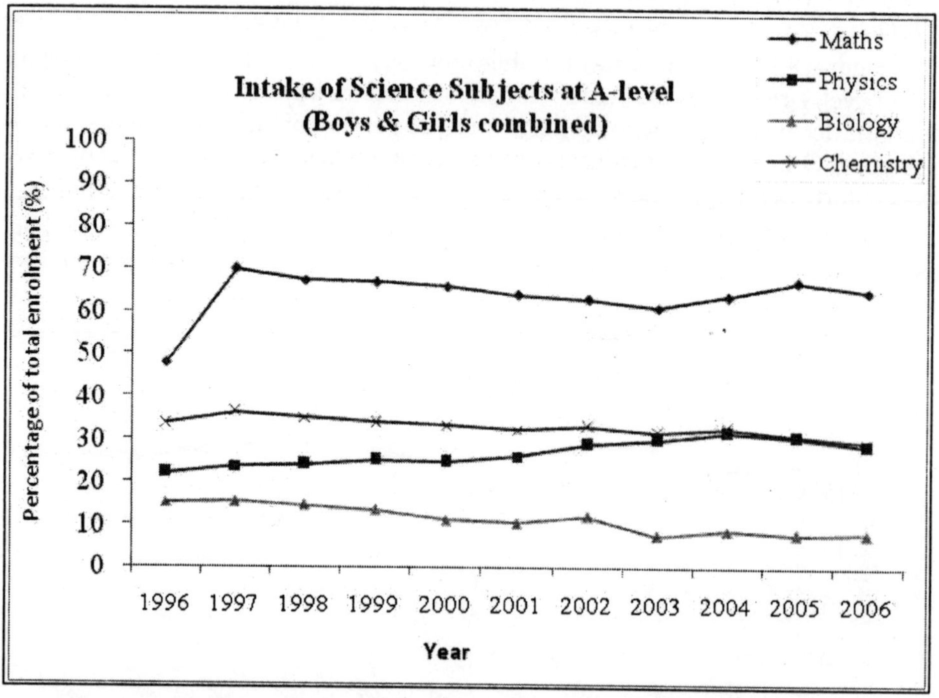

Figure 8.3: Students Taking Mathematics and Science Subjects at A-Level

highlights the decline in intake of the other science subjects. Less than 40 per cent of the students opt for the core sciences.

The status of the access to Information Technology (IT) and the use of Information Communication Technology (ICT) tools in Mauritius are also on the low side and urgent measures should be taken to reverse this trend if the objective is to be a cyber-island in the region.

A survey conducted by the Mauritius Research Council (MRC) (Suddhoo, 2007) highlights the following:

- Some 62.9 per cent of persons aged 12 years and above did not have any knowledge on IT.
- The percentage of persons aged 12 years and above who used the Internet from any place was 18.0 per cent .
- Households owning a computer represented 24.2 per cent of all households while those having internet access at home were 16.6 per cent
- The percentage of primary schools providing Internet access to students for study purposes was 4.8 per cent
- The percentage of secondary schools providing Internet access to students was 92.1 per cent

Following the gloomy image projected by the above observations concerning students opting for science subjects, the Ministry of Education and Human Resources is taking some bold decisions to be in-line with the government policy (Government Programme, 2005-2010). This includes carrying out fundamental reforms in the education system with a view to providing World Class Quality education to enable young Mauritians to be employable in new sectors of the economy, to have more fulfilling jobs and also to be competitive at the international level (Government Programme, 2005-2010). These measures are to further ensure that teaching of science and technology is given a major boost at all levels so that Mauritians can participate fully in a knowledge-based economy and in research and development programmes in various sectors of the economy.

Some of the actions being taken by the Ministry of Education and Human Resources (Gokhool, 2007) include:

- The creation of a Science and Technology Network to ensure better coordination and synergy among local institutions involved in the promotion of science and technology.
- Construction of Science Blocks at several secondary schools. These are buildings specially designed to accommodate science laboratories with high quality laboratory equipment and a proper environment to make teaching of science conducive to learning. This initiative is supplemented with appropriate training for the human capital working in these blocks.
- At the primary level, the School IT Project has made it possible for all schools to be equipped with a fully-fledged computer laboratory. Accordingly, the Ministry is ensuring that there is a rise in the number of ICT teachers and that they receive adequate training to ensure quality education.
- A major achievement has been the launching of the Knowledge Channel on the national television, the Mauritius Broadcasting Corporation (MBC). This channel is supposed to enhance the learning experiences of children and adults alike. Through their manifold presentations, the educational programmes will not only connect students to the outer world but also thrill them to the joy of learning.

Other projects are in the pipeline. Spearheading the willingness to boost the study of science and technology in 2009, a pilot project aimed at broadening the curricular base at Form V will be launched and will be known as the 21st Century Science. While on the one hand this initiative will provide students up to the age of 16 with a varied curriculum, on the other hand, it will contribute to align our education system with international trends, in line with the recommendations of the Curriculum Document of 2006 of the Ministry. In this context, students up to Form V or equivalent will do all subjects, with full specialization occurring only as from the A-Level.

It is also worth noting that the tertiary institutions are increasing their access to Mauritian students to Science, Agriculture, IT and Engineering degree/diploma courses. For example, the University of Mauritius has observed an increase in intake in Science and Engineering degree courses, and the range of courses offered at both undergraduate and postgraduate levels are expanding as per the needs of the country.

While these educational reforms are being implemented, parallel schemes to enhance the awareness of Science and Technology among the public at large must be carried out. This becomes the responsibility of the first science centre in Mauritius, namely the Rajiv Gandhi Science Centre.

3. The Rajiv Gandhi Science Centre

The proposal for the setting up of a Science Centre in Mauritius initiated during the visit of the then Minister of Education and Science[3] to India in 1992. The Science Centre project was approved at the VIIth session of the Indo-Mauritian Cultural Exchange programme in January 1994, whereby the Government of India agreed to assist Mauritius in this endeavour. The Rajiv Gandhi Science Centre Trust Fund (RGSCTF) was then established in 1994 by an Act of Parliament with the aim of promoting Science and Technology, supplement school education through non-formal programmes and create awareness on developments in Science and Technology among the public (RGSCTF Act, 2004). It was named after the Ex-Prime Minister of India, late Mr Rajiv Gandhi who was himself a scientist and geared India towards the present scientific status it owns.

A Memorandum of Understanding (MoU) was signed by the Ministry of Education and Scientific Research and the National Council of Science Museums (NCSM), India, in October 1998, for the setting up of the first Science Centre in Mauritius called the Rajiv Gandhi Science Centre (RGSC). As per this MoU, the Government of India through the NCSM has provided exhibits for five galleries, an outdoor science park, a Mobile Science Exhibition and an inflatable planetarium. NCSM has been closely involved in the planning of the Science Centre and interacting on technical matters relating to the construction of the building. In fact, the first Director of the centre was deputed from NCSM for a period of three years. The Government of

3 It may be noted that the Ministry of Education has been changing its appellation whenever there has been a change in Government or cabinet reshuffle. That is why in this paper it is referred to as the Ministry of Education and Science, or the Ministry of Education and Scientific Research, or still the Ministry of Education and Human Resources.

Mauritius has provided funds for construction and gives grant to Rajiv Gandhi Science Centre Trust Fund for running of the Centre.

The core staff of the Centre was also trained by NCSM. The organisation chart for the Rajiv Gandhi Science Centre is relatively simple. The centre consists of four main Sections (Art, Education, Mechanical Engineering and Electrical/Electronics/IT Engineering) excluding the Administrative and Management sections. There are an established number of 40 officers in the centre out of which about fifty per cent is still vacant.

The Rajiv Gandhi Science Centre was inaugurated on 30 November 2004 by Mrs. Sonia Gandhi, the wife of Late Rajiv Gandhi, and since then, the Centre has received more than 100,000 visitors. Operating under the aegis of the Ministry of Education and Human Resources, RGSC has as one of its main objectives, to inculcate scientific awareness in current and relevant fields and arouse curiosity among the public in general.

4. Status of the Rajiv Gandhi Science Centre

The importance to promote science among the public at large was already felt more than a decade ago, and even during the construction of this unique Science Centre in the region, the NCSM donated a Mobile Science Exhibition named the Museobus. Spearheading our outreach programme, the Museobus is a specially designed bus equipped with 24 interactive science exhibits on the theme "We Are One". The exhibits illustrate the workings of the different organs inside the human body. The scientific facts causing apparent physical differences among respective individuals are highlighted. The message conveyed by the exhibition is that regardless of race, caste or colour, the organs in all human bodies perform the same functions. This mobile bus has visited most of the secondary schools of the island (Oree, 2004).

During its short existence, the Rajiv Gandhi Science Centre has been employing a number of strategies to promote science to the Mauritian population. These tactics have set the pace for the centre to realise its vision of becoming a Centre of excellence in the communication and promotion of Science and Technology in Mauritius.

In the sub-sections that follow the various measures presently employed by the centre are briefly described.

4.1 Galleries

The Centre has five permanent galleries of interactive exhibits and one temporary exhibition hall. Outside the building, visitors can interact with the 40 exhibits in the Science Park to explore concepts of science. The five galleries are described below:

4.1.1 Origin of Mauritius

This gallery explains, through its 15 exhibits, the evolution of the Universe, the Solar System, The Earth and Mauritius. Visitors learn about the different myths on the creation of the Universe, the expanding nature of the Universe, the evolution of the Solar System and of the Earth. In addition, a section highlights the scientific explanation of the origin of Mauritius.

4.1.2 Land and Environment of Mauritius

With its 18 exhibits occupying an area of about 400 m^2, this gallery explores the geology of Mauritius and informs visitors about the important issues regarding the land and environment of our country. Interactive exhibits highlight the fascinating marine ecosystem surrounding our island. The central exhibit portrays the extinction of the Dodo due to irrational exploitation. Information about the weather is available in real-time at the Weather Station.

4.1.3 Resources of Mauritius

The 16 exhibits of this gallery deal with the main resources of Mauritius. Through large 3-D/2-D models, video projections, illuminated maps and displays, visitors learn about the agricultural, textile, energy, fishery and tourism industries among others. A demonstration corner shows traditional crafts of Mauritius.

4.1.4 Frontiers of Modern Technology

As the name implies, this gallery with its 32 exhibits is a gateway to the frontiers of modern technology. Technological progress made in the fields of medicine, biotechnology, dairy, textile and marine farming is explained through participatory exhibits. Other exhibits unravel the mysteries behind LASER, nuclear power, satellite communication, space missions, rockets, etc.

4.1.5 Fun Science

This gallery is an invitation for visitors to understand concepts in science through hands-on experience. Indeed, the 40 exhibits encourage visitors to touch, push, pull, and turn so as to understand scientific principles involved in various subjects such as light, magnetism, electricity, mechanics and sound.

4.1.6 Temporary Exhibition

The first temporary developed by the centre was on the theme "Mauritius through Maps". The second one entitled "Mathematics in our Daily Life" was launched on the occasion of the second anniversary of the opening of the Centre. This exhibition was developed in collaboration with Mathematicians from the University of Mauritius. It takes you on the journey of the importance of Mathematics in all aspects of life. Aspects covered include History of Mathematics, Art, Games, Finance, Manufacturing, Cancer Research, and Google Search. The third temporary exhibition which will be developed this year will be on "Inventions and Discoveries that changed our life." In view of the new concept of "Maurice Ile Durable" or Sustainable Development for Mauritius, RGSC is planning a national exhibition entitled "Renewable Energy for Sustainable Development for Mauritius" in 2009.

4.1.7 Science Park

In the outdoor Park, action-oriented exhibits that aesthetically merge with the colourful landscape provide a unique playground where education merges with fun and entertainment. Thus, in some of the exhibits of the Science Park, visitors swing to learn the principles of a pendulum, pull weights on inclined planes to note that the angle of inclination eases man's effort in doing work, sit and turn to find out why we see only one face of the Moon, roll balls to explore the principles of an eclipse, come face to face with dinosaur models.

4.2 National Competitions

It has been shown that a way to motivate the young minds and enhance scientific skills and creativity is through competitions. In this context a range of competitions has been organised for secondary level students since 2005. This year a new competition is being launched for upper primary schools entitled Junior Science Adventure (JSA). It aims at creating an interest in science among young pupils and to initiate them to the investigation of contextual science problems encountered in their daily life. The theme for JSA 2008 is "Me and my Environment."

4.2.1 National Science Challenge

This is a project-based contest, aimed at science students of upper secondary schools (Form IV and Lower VI). Students are expected to undertake projects in the following subject categories: Physics, Chemistry, Biology, Mathematics, Environment and Technology, and take part in individual or group categories. Previous experience shows that this competition has not only revealed the hidden talents of our teenagers, but it has also served as a platform for the sharing of ideas among students from different backgrounds and schools. A high-level and independent panel of jury assesses the projects through various stages and the most outstanding project is awarded the Young Scientist Award. Such a project has to be original and innovative and the student/s should be able to master the scientific aspects of the project. The Young Scientist Award for the year 2007 was attributed for a project on "The Bamboo Ecological House."

4.2.2 Science Communication Contest

This competition is aimed at lower secondary school students (Forms 1-3). The main aim of the competition is to encourage young students to research on a scientific topic, fully understand it and then prepare a Microsoft PowerPoint presentation to be delivered in front of an audience so as to encourage/train our young students to tap their inner potentials and communicate effectively. The preliminaries are held in each of the four zones and the final takes place at RGSC. It is amazing to observe how young minds choose topical issues such as global warming and are so eloquent in their deliveries.

4.2.3 Science Quiz Competition

The contest has been designed for various secondary level students (Form 3 and Form V) in the past and has even been jointly organised with the Mauritius College of the Air. A screening test is held to select 16 best teams for the competition. The various stages include preliminaries, quarter finals, semi-finals and then the Final which is live. Each team consists of two students and the questions cover curriculum-based subjects such as Physics, Chemistry, Mathematics, Biology and Life Sciences, as well as general knowledge subjects which include Astronomy, Inventions and Discoveries, Information and Communication Technology, Earth Sciences and Environment.

4.3 Educational Activities

Since the colonization of Mauritius by the British, the education system provided to our youth has been highly competitive and examination oriented. Until more recently this situation has deteriorated with the proliferation of private tuition in the community. This present system does not encourage extra-curricular activities and hence suppress creativity among children. Yet, there is a natural instinct in the young ones to explore newer vistas to satisfy their curiosity. When an element of play is coupled to this exploration, it allows children to be more creative and helps in the development of their linguistic, cognitive and social skills. Considering all such aspects, activities mainly aimed at developing creative faculties in children have been regularly chalked out by the RGSC.

4.3.1 Workshops, Seminars and Popularisation Lectures

A regular feature of the centre is the organisation of the Young Scientist's Workshop which is held thrice annually during the school vacations, and is aimed at the lower secondary school students. Each holiday a new scientific theme is presented and the children get a manual and one-to-one attention for hands-on experiments. They are exposed to lectures, model-making sessions and science games. Due to high demands, more sessions are offered on the same topic and recently the number of intakes has increased by 50 per cent .

With the view to inculcate new techniques and develop low-cost environment friendly experiments to teachers, Teachers' Science Workshops have been tailor-made and held during school vacations. Special workshops have also been conducted for teachers following the setting up of science laboratories also known as Science Labs in some primary schools of the island. The Rajiv Gandhi Science Centre has been strongly supporting the government in the setting up of Science Labs in primary schools by providing training to teachers and technical support whenever required.

Besides designing activities for school students, several lectures have been delivered by local and visiting scientists on a wide range of subjects to the general public. These Popular Science Lectures, as they are called, can also be associated with major events such as "The Transit of Venus" in 2004.

With a parallel objective of going in-depth on certain topics, one day seminars are organised for the public at large. The last seminar entitled "Threat from Volcanoes in the Region–Is it Real?" was held in September 2007 and was a huge success.

4.3.2 Sky Observations

Sky observation programmes are scheduled bimonthly for the public to gaze at the moon and planets of the solar system through the telescopes of the centre. In this activity, visitors are provided with a sky chart. They are explained how it should be used to locate different constellations with the naked eye, their time of visibility as well as their positions in the night sky. At the end of the session, the many and varied questions regarding astronomy are answered. As an integral part of the Sky Observation, lectures on astronomy with the help of slides or PowerPoint presentation are also delivered. Many subjects pertaining to astronomy are tackled: the night sky, zodiacal constellations, the phases of the Moon, the planets of the Solar System, why and how the constellations change positions in the sky at different times, birth of stars, galaxies and the universe. The RGSC has four telescopes of Celestron make. The visitors are given the opportunity to view craters of the moon, planets such as Venus, Mars, Jupiter and Saturn through telescopes. Faint objects such as nebulae and galaxies are also observed.

4.3.3 Planetarium Shows

It has been a tradition that the Inflatable Planetarium accompanies the Museobus, mentioned in Section 4.1, to introduce the students to astronomy as well. The portable planetarium is used to initiate students to the wonderful world of astronomy. Presently, the show which is most appreciated by students is the night watch. In this show, the participants get a stimulating view of the night sky. The different constellations, their time of visibility as well as their positions in the night sky are pinpointed. At the end of the session, the many and varied questions regarding astronomy are answered.

4.4 Other Facilities

Located at the South entrance of the capital city, Port-Louis, in Bell Village, at the foot of the Signal Mountain, the Rajiv Gandhi Science Centre is the ideal venue to hold lectures and conferences, and related activities, with ample parking facilities. The fully equipped auditorium is frequently booked by various private and government institutions for special functions.

The library is constantly improving its stock and is aiming to become the National Science Resource Centre. Other facilities include a cafeteria and a souvenir shop which are not yet fully operational. However the cafeteria is always used for catering purposes during functions.

5. The Way Forward

The Rajiv Gandhi Science Centre has a crucial role to play in the promotion of Science and Technology. In order to keep on improving its image locally as well as internationally it has to be able to dynamically adapt to changing situations by providing tailor-made programmes for specific segments of the population as well as organisations. Besides the varied strategies already implemented by the centre, a series of new measures are being instigated either on its own or in collaboration with other institutions.

5.1 Science and Society

At the dawn of the 21st century, science and technology (S&T) are playing an increasingly important role in determining the destiny of countries around the world. For developing countries like Mauritius, modern S&T are even more crucial as they offer hope for addressing the pressing needs of the population in terms of improved nutrition, public health, safety, and shelter. We are on the brink of exciting developments in S&T that will affect everyone's lives; and along with the benefits come potential risks which cause understandable concern. People rightly expect proper safeguards against risks, and full and open discussion about the values, benefits and risks of scientific developments. In today's knowledge-based society, government must ensure that citizens are able to make an informed choice from the options made available to them by responsible scientific and technological progress.

Examples abound in Mauritius to illustrate how the widespread support and participation of the public is vital for the implementation of policies: from AIDS to global climate change, from sustainable energy to forensic science, from ICT to Biotechnology. By encouraging and enabling society to take an active part in the debate about the science and its applications, we could ensure that all our citizens get the opportunity, not only to contribute their skills and talents to enrich of our country, but also to participate meaningfully in deciding the future course for Mauritius. Failure to respond effectively to these challenges will reflect negatively on our economy, and lead to stagnation in a time when S&T are the growing forces of healthy economies.

The Rajiv Gandhi Science Centre (RGSC) has an important role to play in taking forward this agenda and engaging the public with science and the issues that it raises for society. RGSC offers a wonderful opportunity for raising public awareness of current scientific matters, providing a neutral venue for activities and programmes, and providing opportunities for students and the public alike to experience the wonders of science. Adequate knowledge and understanding of science will strengthen the confidence and self-reliance among our people. This is the only way to maximise public participation in science and technology policy making and delivery processes.

To this end, RGSC is developing an action plan aiming at the promotion of science in the Mauritian society through a wide variety of societal added-value activities.

5.1.1 Science Outreach Programmes

They have been redesigned to target, at an initial stage, students of primary and secondary schools around the island. Besides the Museobus and the Planetarium, which will continue to form part of this scheme, Science Talks and Science Shows have also been included.

The Science Talks aim at increasing the awareness of students in different science subjects, arousing a scientific thinking and an inquisitive mind among students and at some level supplementing their school curriculum. The topics proposed by RGSC in the first instance are: Simple Machines, Earth-Moon System, History of Computers, Climate Change, and Astronomy.

Science Shows, alternatively named "Science Circus", will be a regular feature in secondary schools and also in primary schools. The Science Shows will be designed for Form 1 to Form 3 students of secondary schools of all zones around the island. Similar activities are being planned for Standard V students of primary schools, in priority those schools which have science activity classrooms or science laboratories. The interactive Science Shows are designed to arouse the students' interest in experiencing and experimenting science and also to motivate and entertain them. The proposed themes for the Science Shows are: Unexpected Science Show, Fluid Show, Pressure Show, and Cold Show using Nitrogen.

These programmes will eventually be conducted for different organisations, village councils etc.

5.1.2 Science Film

Shows will be conducted on topical issues of S&T in view of raising public awareness. The selected films and documentaries will delve into matters that are of current interest at the national level. Suggested themes are global climate change, diabetes and hypertension, sustainable development, urban transport, eclipses and astronomical phenomena, *etc.* While these shows will be hosted by the RGSC auditorium, they can also be held at schools, community centres, district councils and municipalities in order to reach out to the public.

5.1.3 National Campaigns

National Campaigns are usually carried out by Ministries, Non-Governmental Organisations (NGOs) and other stakeholders to promote public understanding of issues of interest for the public at large, which often are scientifically relevant. Recently, two such campaigns have been launched: the Energy Savings Campaign and the AIDs campaign. The RGSC, being an expert in the field of design and development of interactive exhibitions, could develop synergies with the stakeholders to help them deliver their message in innovative and attractive ways.

5.1.4 Commemorative Events

Commemorative Events are observed every year to celebrate special occasions. Examples are the World Water Day, World Day of Telecommunications and the International Day for the Preservation of the Ozone Layer. RGSC intends to collaborate with relevant organisations, public authorities and the media to disseminate relevant information (in the form of seminars, lectures or discussions) to the civil society so that the public is better informed about the importance of the issues.

5.1.5 Science Awareness Programmes for Housekeepers and the Elderly

If society as a whole is to better understand and accept the developments in science and technology, measures must be taken to address the under-representation of specific categories of the society in science in Mauritius. In this context, special attention should be paid to housewives and the elderly. Both housekeepers and the elderly form integral parts of society but are often overlooked when it comes to S&T although science and its applications are intrinsically present in their everyday lives. Thus, RGSC can design tailor-made programmes for them in collaboration with the Ministry of Women's Rights, Child Development and Family Welfare and the Ministry of Social Security. The programmes would aim to enhance awareness of housewives and elderly about matters for which they are directly and indirectly involved including safety aspects at home and health basics.

5.1.6 Sky Observation Programmes

In view of increasing access to our telescopes and other resources for astronomy, it is proposed to organise night sky observation programmes in villages and towns with the collaboration of district councils and municipalities.

5.1.7 Popularisation of Science through the Media

The media is a powerful means to popularise S&T in Mauritius. RGSC proposes to develop fruitful collaborations with the media, both audiovisual and written, to foster public understanding of science concepts and issues. Thus, RGSC can contribute simple articles in newspapers to demystify scientific phenomena which are not readily understandable to the public. RGSC can also participate in discussions on scientific issues on the radio and TV. In addition, RGSC offers to launch a quarterly publication RGSC Science Bulletin that will contain practical and interesting information on S&T for the public in general and the student community in particular.

5.1.8 Science Workshops

The Rajiv Gandhi Science Centre is proposing to organize a series of workshops related to Science and Society. This activity aims at acquiring and disseminating the latest information in science and technology and also creating awareness of the impact of Science and Technology in society. It endeavours to stimulate interest about issues that concern people and to connect these issues to science. Thus RGSC will be helping in creating a scientifically and technologically literate society.

A general public with an understanding of the nature of science produces a skilled workforce able to compete in a knowledge-based global economy; individuals capable of making more informed decisions and prepared to engage in matters involving science.

The Rajiv Gandhi Science Centre will collaborate with likeminded institutions and various Ministries to develop a series of workshops that brings together practitioners of science committed to the cause of enhancing understanding of Science and Technology and the broad spectrum of the society.

5.1.9 Development of Exhibitions

Promotion of Science and Technology through interactive exhibitions no doubt remains the core activity of the Rajiv Gandhi Science Centre. After being fully operational since December 2004, the visitors to the Centre have been mainly students. Exhibitions have to be upgraded and new ones developed for the general public and foreign visitors. In line with the strategies to achieve the vision of RGSC, exhibitions in the field of Energy, Global Warming, Information Technology and Biotechnology can be developed. This is possible through local and international collaborations.

While developing new exhibitions older ones will have to be dismantled due to space constraint in the Centre. It is proposed to move these exhibitions travelling around the country in schools or other institutions.

5.1.10 Development of a National Science Resource Centre

The Resource Centre proposed at the Rajiv Gandhi Science Centre will target all layers of the population. Facilities to be offered include: access to internet, interactive VCD and DVD viewing and documentaries on a range of scientific topics such as astronomy, animals plants, oceans/marine life, technology, physical sciences, *etc.* The resource centre will also be a source of scientific information in hard and soft formats for assignments and essays for students of all age group and the public in general.

5.1.11 Networking of RGSC

It is proposed that Internet facilities be offered at several different points in the permanent galleries of the centre. The networking project of RGSC will take care of this aspect of the galleries. This facility can be used by visitors to look for further information on any specific exhibit of their interest and it can also provide a database of links to different sites of scientific relevance. One of the theatres will be equipped to host Video Conferences and participate in major regional and international such as launching of rockets, debates on climate change, global warming, desertification, lectures by eminent scientists, etc. Depending on the level of the debates, students, the general public or local scientists could be invited to take maximum advantage of this opportunity.

5.2 National Science Website

The Ministry of Education and Human Resources has set up a Science and Technology Network in 2007 (refer to Section 2) to pool all institutions involved directly in the popularisation, education and research in S&T. One of the proposals of this committee was the setting up of the Mauritius Science Portal[4] and RGSC is a major player in this project.

The main objectives of this website are:

- To group all stakeholders in Mauritius directly or indirectly connected to Science Education in Mauritius,
- To be an Educational platform for Scientific knowledge,
- To be the science information centre for all layers of the Mauritian population,
- To provide a database of local and international resources in Science.
- To be a one-stop shop for S&T in Mauritius

It may be highlighted that RGSC will be responsible for the chapter on the promotion of S&T in Mauritius.

5.3 Multimedia Development

Following a proposal from the Science and Technology Network, the Rajiv Gandhi Science Centre shall collaborate with appropriate stakeholders to develop a first set of CD for Form 1 on a pilot basis with the aim to: to supplement the existing syllabus; to reinforce the understanding of basic concepts in science; to encourage the use of ICT tools; to make teaching and learning of science more lively; and, to arouse interest in science.

5.4 Knowledge Channel

The Mauritius Broadcasting Corporation, which is the national television in Mauritius, has recently launched the Knowledge Channel in September 2007 with a view to educate the population through various television programmes. The Ministry

---

4 The Mauritius Science Portal has been officially launched on 30 March 2009.

of Education and Human resources has been allotted four hours of media time out of which half an hour will be devoted to science related programmes to be developed by RGSC. The Rajiv Gandhi Science Centre has come up with an action plan in this direction to satisfy all layers of the population. Some of the programmes proposed are:

- Discovering Science
- RGSC Forum
- Science Movies
- Science Interviews
- Science in Schools
- Frontiers of Technology in Mauritius
- Trades in Science and Technology

All the programmes are being developed in collaboration with the Mauritius College of the Air, and a few have already been aired on the national television.

5.5 Local and Regional Collaborations

Due to the resource constraints of the Science Centre and the increasingly important role of Science and Technology in the socio-economic development of Mauritius, it has become imperative to seek collaboration from local, regional and international organisations to develop programmes to meet the stringent demands of the population. As mentioned earlier, through the Science and Technology Network, the local institutions are already pooled together and interesting outcomes are already being visible.

Regional collaboration is also moving in the right direction. As mentioned earlier, in September 2007, the RGSC conducted a national one-day seminar entitled "Threat from volcanoes in the region–Is it real?" This event was blessed by the presence of the Director of the "Observatoire Volcanologique de Piton de La Fournaise", in Reunion Island, and was a huge success on all fronts. Some recommendations to be approved by the Government have even emerged from this seminar.

As stipulated in the introduction of this paper, RGSC owes its existence to the National Council of Science Museums, India. The Memorandum of Understanding signed between the Government of India and the Government of Mauritius in 1998, as mentioned in Section 3, lapsed in 2006 following the departure of the previous director. In the 10th Session of the Indo-Mauritian Joint Commission held in Mauritius in December 2007, the avenue for the further development of RGSC opened. The following were agreed in principle pending the signing of another MoU:

- Construction of a 120-seater Hybrid Planetarium in the premises of RGSC
- Extension of the RGSC building for the development of two new galleries
- Development of new exhibits for the centre, and,
- Training of the core staff of RGSC

This new development will definitely be a milestone for the Rajiv Gandhi Science Centre. Besides improving RGSC's financial situation, a Hybrid Planetarium will also add a regional dimension to our educational and tourism sectors.

6. Conclusion

This paper has browsed through the various milestones of the first science centre in Mauritius. Starting from the present status of Science and Technology in Mauritius, the various measures initiated by the Government of Mauritius to reverse the deteriorating image of science among the students and the public at large are highlighted. The challenging role of the Rajiv Gandhi Science Centre to promote science among the population is clearly shown. Starting from the galleries, through the present activities designed for the young generation, the way forward is clearly drawn. The Rajiv Gandhi Science Centre is fully prepared to support Mauritius and its neighbours in this scientific and information era.

It may be highlighted that since the cabinet reshuffle in September 2008, the Rajiv Gandhi Science Centre operates under the new Ministry of Industry, Science and Research and hence it has a major role to play in the economic and social development of the island. In fact, through the initiatives undertaken, the Rajiv Gandhi Science Centre shall continuously contribute indirectly to the development of Mauritius.

Acknowledgements

The support of the Ministry of Education and Human Resource is placed on records for my nomination to the 5th International Workshop on Enhancing Change through Science Centres. I am extremely grateful to our Honourable Minister Mr. D. Gokhool for his support to the centre and for his vision for S&T in Mauritius. I wish also like to thank all the Staff of the Rajiv Gandhi Science Centre for working as a family to give the centre its present national dimension it owns.

References

Gokhool D., 2007. Speech on the occasion of the Launching ceremony of the Science Festival 2007, Rajiv Gandhi Science Centre, Mauritius.

Government Programme, Mauritius, 2005/2010.

http://tourism.gov.mu (accessed 16 January 2008)

MRC,2004. Teaching and Learning of science in schools in Mauritius, Volume I, Recommendations and Action Plan.

Oree, V., 2004. "Current Status of Science Centres and Museums in Mauritius." Second International Workshop on "The Changing Role of Science Centres and Museums in the Developing Countries," Vietnam.

RGSCTF Act, 2004. Rajiv Gandhi Science Centre Trust Fund Act 6 of 1994. Revised Laws of Mauritius 2000, Volume 6.

Selvon, S., 2001. A Comprehensive History of Mauritius. Mauritius Printing Specialists.

Suddhoo, A., 2007. "Science and Technology in a small island developing state: The case of Mauritius." Science Colloquium, Rajiv Gandhi Science Centre, Mauritius.

Abbreviations

- ICT: Information and Communication Technology
- IT: Information Technology
- JSA: Junior Science Adventure
- MoU: Memorandum of Understanding
- MRC: Mauritius Research Council
- NCSM: National Council of Science Museums
- RGSC: Rajiv Gandhi Science Centre
- RGSCTF: Rajiv Gandhi Science Centre Trust Fund
- S&T: Science and Technology

Chapter 9

Science Awareness Campaigns

Syed Azhar Hasan
*Pakistan Science Foundation, Constitution Avenue,
Islamabad, Pakistan
E-mail: syedazharhasan@hotmail.com*

ABSTRACT

Popularization of science plays a central role in the socioeconomic, cultural, and environmental development of any country. In socioeconomic terms, the popularization of science encourages talent for scientific research, technological development, and intellectual endeavors. Popularization of science is one of the statutory functions of Pakistan Science Foundation. The Foundation is engaged in various science popularization activities at national level with the aim of increasing awareness of science education in the society. In order to achieve these objectives, the Foundation has initiated a number of programmes including science exhibitions, science fairs, science film shows, popular science lectures and science quiz competitions, science essay and poster contests, establishing of science museum/centres/corners, strengthening of laboratories of high schools etc. The Foundation has already established a Natural History Museum, six Science Centres and nine Science Caravans (Mobile S&T Exhibitions). These are ideally poised to play an important role in nation-building efforts through innovative science exhibits which stress aspects of everyday life and other developments relevant to a nation. The Foundation has also organized Science and Technology Fairs/Expos wherein the main focus was on popularization of Science among students and general public. The Foundation has now planned to establish a state of the art Science and Technology Museum at Islamabad, Science and Technology Corner at Pakistan Monument Museum and a network of 15 Science Centres and 500 Science Clubs in the country.

Keywords: *Pakistan, Science popularization, Science Caravans, Science Centres, Science Museum, Science Clubs.*

Introduction

Science and Technology (S&T) is a vehicle for economic growth and for improving the quality of life and superiority of a nation. It provides humanity with the means to grow through applicable and implementable solutions for complex problems, and has served continually as an instrument for building the prosperous world. Nonetheless, science has also contributed to the current environmental social and economic predicaments faced by humanity in the 21st century.

The developing countries like Pakistan have lagged behind in modern science and technological progress. The problem therefore with us is of different nature and of different magnitude. It is more due to lack of public understanding of science and assimilation of technological development by masses for their own benefits in the society rather than ill-effects of scientific advancements vis-à-vis image of scientists and technologists. Until and unless science is not ingrained in society, the benefits of S&T development are not appreciated by the masses and the culture is not transformed into a scientific culture; the dream of science and technological development leading to self reliance in the country simply cannot be realized (Hasan, 2002; Kamble, 2006; Mufti, 1997).

Popularization of science plays a central role in the socio-economic, cultural, and environmental development of any country. In socio-economic terms, the popularization of science encourages talent for scientific research, technological development, and intellectual endeavors in general. It fosters creativity and innovation, contributes to producing better trained human resources, expands social opportunities, and strengthens the educational system. Culturally and environmentally, the popularization of science enhances the critical sense of the population, by increasing its involvement in decision-making and contributing to sustainable development.

Science Popularization Initiatives

Pakistan Science Foundation (PSF) established in 1973 as a premier organization for promotion and popularization of science in the country. The Foundation is engaged in various science popularization activities at national level with the aim of increasing awareness of science education in the society. In order to achieve this objective, the Foundation has initiated a number of programs including science exhibitions, science fairs, science film shows, popular science lectures and science quiz competitions, science essay and poster contests, establishing of science centres/corners, strengthening of laboratories of high schools, etc.

Science Museum/Centres/Corners

The Foundation has already established a Natural History Museum, six Science Centres, three Science Corners and nine Science Caravans (Mobile S&T Exhibitions). These are ideally poised to play an important role in nation-building efforts through innovative science exhibits which stress aspects of everyday life and other developments relevant to a nation. These are important in discovering and exploring ideas and also in helping to forge a strong relationship between science and society.

The role of Science Museums and Science Centres has also expanded as they are not merely concerned with the procurement, care, study, display of objects but also are the information centres for understanding our world, raising public awareness of current research and stimulating an interest in science among people. PSF also offers help in the form of exhibits and expert guidance to educational institutes interested in setting up Science Corners on their own.

Science Caravans (Mobile Science Exhibitions)

Science Caravan is a Mobile Exhibition that has been launched to popularize science and create science awareness among people living in rural/backward areas of the country. The Science Caravan makes tours of the schools and even colleges in the rural areas and sets up the exhibition for a week or more in one of the school. These science exhibitions are visited by students of neighbouring schools as well as rural population. The visits to various schools are arranged in collaboration with the relevant Directorate of Education.

The Science Caravans have four major parts:

Panel Exhibitions

Each Science Caravan has a set of twenty panels consisting of large scale photographs and write-ups (both in Urdu and English) on such diverse themes as lasers, medical break-through, space exploration, renewable energy resources, nuclear energy, mineral wealth of Pakistan, environment, biodiversity, etc.

Equipment and Working Models

Mechanical models, Scientific Exhibits and three-dimensional displays are additional attraction of the Caravan exhibitions. Most of the models are related to themes as per syllabus of the SSC or HSSC level.

Starlab (Copyright) Planetarium Show

Planetarium Show is the most attractive and interesting part for all categories of the students.

Science Film Shows

Science Film Shows are another important part of the Science Caravan Exhibitions. Each Caravan has 16 mm Film projectors along with 16mm films, mainly produced by Encyclopedia Britannica and Coronet Corporation. PSF has now provided Multimedia Projectors and Projection Screens to all Caravan units. The original films imported from the producers were in English, but keeping in view the difficulties faced by the students of rural area, PSF has dubbed the films in Urdu language. The films have also been converted on VHS as well as on CDs, which have been provided to the Caravan units. Major themes of the scientific films are Energy, Environment, Universe, Medical Sciences, Robotics, Computer Science, Geology, Sea Life, Jungles, Electricity, etc.

PSF has planned to use the Science Caravans as Community Science Centres in the evening, which will propagate Science in the general public and out of schools children.

National S&T Fairs and Expos

The Foundation has been organizing Science Fairs/Exhibitions from time to time in the country. These Science Exhibitions act as a vehicle for propagation and promotion of science and technology. Not only the young students get inspiration from our scientific developments and from indigenous products of high technology comparable to the products of advanced countries.

Scientific fascination also invites them to adopt careers of science in their future lives. The information on these scientific and technological developments also invokes in them the confidence and they get imbibed with the spirit of making such contributions to their homeland in times to come wherein large number of R&D/S&T organizations, Entrepreneurs and Industrialists participate. S&T Expo-2007 was organized at National Monument, Islamabad from 24th March to 23rd April, 2007. The Expo highlighted the achievements of the public sector Research Organizations and to give the general public, particularly the younger generation, an opportunity to observe the major achievements in Science and Technology.

Strengthening of the Laboratories of the Govt. High Schools of Rural Areas

Since most of the high schools working in government sector in the rural areas are lacking in laboratory equipment, PSF in consultation with respective Directorates of Education provides laboratory equipment to deserving high schools to strengthen their laboratories. Computers have also been provided to some schools.

Preparation of Science Posters for Schools

Preparation of science posters, brochures and leaflets on important and interesting scientific themes and concepts is another important programme of the Foundation. More than 10,000 copies of each poster and other reading material are being printed and distributed to schools.

Science Poster, Essay and Quiz Competitions

Science Essay and Poster Competitions are very successful programmes of the Foundation. PSF in collaboration with 22 Boards of Intermediate and Secondary Education organize the competitions among the students on important scientific topics. Sixteen Science Essay and Poster Competitions have been organized so far. Winner students are awarded with medals and cash prizes.

Popular Science Lectures

Eminent scientists, technologists and educationists give lectures in non-technical language for the mixed audience, comprising scientists, research scholars, students and the general public. The lectures are held frequently and widely announced in the press.

Collaboration with International Agencies

PSF also collaborates with international scientific bodies like the United Nations Educational, Scientific and Cultural Organization (UNESCO). PSF is a member of

UNESCO Sub-Regional Network of Non-Formal Science Education and Popularization of Science. The Foundation is involved in UNESCO project "Breaking the Poverty Cycle of Women" since June 2002. Supplementary reading material, science kits, posters and brochures on various scientific themes have been prepared by PSF (especially for adolescent girls) for the neo-literates of rural areas of Pakistan.

Special Events

Celebration of special events is another important activity of the Foundation. World Science Day for Peace and Development (WSDPD) on *November 10, of each year,* has become a regular programme of the Foundation. World Science Day was proclaimed by UNESCO in the year 2001, since then the Foundation in collaboration with Ministry of Science and Technology and UNESCO is holding various activities like, Convention of Scientists, Speech Competition among students and Science Exhibitions.

World Year of Physics-2005 (WYP) was also celebrated by the Foundation by organizing lecture for students on the occasion, "Einstein-The Genius of The Century".

Live Talk Shows

PSF frequently, organizes live Talk Shows on PTV World and other TV channels, coinciding with major science popularization activities like celebrations of World Year of Physics-2005, World Science Day for Peace and Development November 10. Eminent national and international scientists and educationists participate in the live Talk Shows.

Science Awareness through Mass Media

Regular science features are contributed by the scientists of PSF to national News Papers, covering a diversity of topics. Research papers are also published in periodicals and journals.

Meeting of Nobel Laureates with Pakistani Students/Young Scholars

Meetings of the students with Nobel Laureates and other scientific celebrities are always very exciting for the young generation (Butt and Butt, 2003-2007). Being inspired by their achievements, the young generation can opt for higher education and research in science. PSF has already conducted two meetings of Nobel Laureates. Purpose of the meetings is to inspire and motivate Pakistani students, young scientists and research scholars for developing scientific culture among students and to broaden the outreach of the country's scientists.

- First Meeting of three Nobel Laureates viz; Prof. Charles H. Townes, Prof. Ivar Giaever, Prof. Gerardus 't Hooft and a German scholar of Nuclear Medicine, Prof. L. E. Feinendegen with the Pakistani students/young scholars was organized by PSF in 2006.
- 2nd Meeting of Nobel Laureates with Pakistani Students/Young Scholars was very successfully organized from 27-31 March 2007 at Islamabad and Lahore by the Foundation in collaboration with PAEC/PIEAS, HEC and MoST. Three Nobel Laureates; Prof. Peter Agre, Prof. Klaus Von Klitzing and Prof. Douglas Osheroff participated.

New Initiatives

Setting up of Science Clubs in High Schools

Establishment of Science Clubs in high schools is a very important and useful programme. PSF intends to establish 500 Science Clubs in the schools by providing them technical and financial assistance. Major objective of the activity is to develop technical skills in the students alongwith developing their interest for carrying out scientific and technological projects.

Establishment of 15 Science Centres

PSF intends to establish 15 more Science Centre in different cities. These science centres will play a pivotal role in propagation and popularization of science and technology, its history and especially its applications for the benefit of mankind and also to make the economy self reliant.

Establishment of a National Science and Technology Museum

At present there is no national level S&T Museum in the country. The Foundation has planned to establish a National Science and Technology Museum at Islamabad. Basic ideas and advanced concepts of science and technology will be displayed through interesting and attractive exhibits and working models in the museum which will be a source of continuous propagation of science and technology among children and the public at large.

References

Butt, N.M. and Butt, W.A. 2003-2007. Five years of Interaction of Pakistani Students with Nobel Prize Winners in Landau (Germany). PIEAS, Islamabad, Pakistan, pp.1-137.

Hasan, S.A., 2002. Natural History Museum: A vital source of biodiversity information. In: Afzal, M., A. Abid and A. Nasim (Eds.) Information handling in Biological Research pp. 49-56.

Kamble, Vinay B., 2006. The Enhanced Role of Science Centres and Museums in Development Strategies: Proc. African Regional Conference, Lusaka, Zambia

Mufti, S.A., 1997. Biodiversity: Awareness and education through natural history museum. In: Mufti, S.A., C.A. Woods and S.A. Hasan (eds.) Biodiversity of Pakistan, pp.13-24.

Chapter 10

Scientific Education in the Vietnamese Museum System

Ngo Thi Loan* and Thai Van Tan**
Ministry of Science and Technology, Hanoi, Vietnam
*E-mail: *ntloan@most.gov.vn, loancrd@yahoo.com, **tvtan@moste.gov.vn*

ABSTRACT

According to official statistics, currently Vietnam has about 115 museums. At the establishment in the late 1950s of the 20 century, besides the role of research, collection, maintenance, museums have determined their main function of public training. Social museum system has the role of training political thoughts, traditional history of the nation, traditional culture of ethnics over historical periods. Natural museum system has trained the public about knowledge in sectoral science through different scientific museums such as the museum of geologic in Pham Ngu Lao street (Hanoi), the museum of biology at National University, the museum of oceanology in Nha Trang or the museum of ecological animal in Dalat. The main education of most museums is undertaken through activities such as picnic, propaganda, dissemination, introduction of available collection to visitors which are seldom presented in scientific subjects and not animated. Knowledge given by guiders to the public is obtained by passive and inexperienced ways, and therefore less visitors come to the museums. Due to high demand of the public, some scientific museums have improvements in accelerating educational function via abundant educational types such as imaginary museum, presentation of collection, scientific subjects to students via internet or special website. The general plan of Vietnamese museum system until 2020 approved by the Prime Minister on 23/6/2005, has planned to improve and develop museum system in order to meet the requirement of researching, studying and training about history, culture of the public and encourage the development of economics and society. One of the content in the plan is the development of special museums about science and technology below the Ministries to help scientific activities and training via the activities of museums.

Keywords*: Vietnam, Scientific education, museums, Knowledge, Training, scientific temper.*

Education Role in Vietnamese Museums System

Since the end of 19th century, Vietnamese museums regarded as an educational institution and played very important role in "unofficial" training public knowledge of social sciences and natural sciences. Education role of majority current museums is mostly expressed in role of educators and supporting individuals in learning and researching.

Role of Educators

This is undertaken through model of display and education programs.

Display and Exhibition

At any museum usually having permanently exhibits, exhibitions as well as exhibits of special subjects, are considered as basic instruments of education. Guiders shall try to transmit knowledge through collection of objects on display contained cultural and scientific values exhibits, whereby visitors could receive latest knowledge, verify and experience with awareness of their own. Exhibitions and displays of special subjects were always held aiming at specific topical targets, it is closed to the goal is training currently issues.

Education Programs

Any museum wanted to be recognized as modernly educational institution, is subject to supply every public education programs, from regularly training to voluntary training, from children to mature ages and from compulsory education to specific education.

Following is some most commonly education programs in Vietnamese museums:

Outdoor Education Program for Junior High School Students

Museums showed interest and paid a special attention into this program, they combine their own program with teaching program at schools. Museums assign their scientific officials to the school for teaching or talking about special subjects or taking students to attend extra–curricular activities at museums.

Family and Children Program

Museums are supplying children plentiful cultural environment, awakening all of their senses and stimulating their arguments and desires to discover mysteries. Parents and persons are responsible for looking after children shall play a positive role in encouraging children to study or taking part into this program with their children. In addition, museums encourage parents to find out new ideas to make sure that the programe provides variety and is attractive. Program is teaching children how to make ceramics, learning about Vietnamese traditional medicines or playing folk games at Vietnamese Folk Museum. These are examples of the combination between learning about knowledge, psychology, emotions and demands of children with objectives of museums as propagation and conservation of traditional. Those programs received broad correspond of the students, handicraftsmen or professional craftsmen in society.

Figure 10.1: Craftsmen at Phu Lang Ceramic Village is Giving Students Instructions to make Ceramics

Figure 10.2: Vietnamese Students are Playing Folk Games with Japanese Students

Figure 10.3: Museum Officials are giving Students Instructions to Make Scientific Experiments

Training at Higher Education for Student

Program is used for students/trainees/postgraduates, supplied by museum, shall give a little time to practice at museum, study collection and working with curators. Students shall be given instructions by the scientific instructors. However, this is meeting with difficulties as museum could not satisfy all students' requirements because some museums without skillful officials to give students their instructions to study.

Role of Supporting Individuals in Learning and Researching

This is undertaken through "oriented self–study", specific activities of this supporting public to access original exhibit collection, information and giving space for public to study within museums. Professional instructions of museum officials shall make source of knowledge of museums is accessible for public. Facilities of the museum for public self–study are including: Museum library, bulletin, training room, discovery room, "virtual" museum (online museum), hand–on collection, detachable models, games, workshop, interactive equipments and audio guide.

General Assessment

In the role of an education institution, Vietnamese museums already and have been contributing significantly to preserving and promotion of traditional cultural heritages, playing important role in traditional education and promotion, encouraging people in classes to contribute for protection and construction of national independence and Homeland unification. Simultaneously, they contributing remarkable to development of science, education, especially social sciences and enhancement of intellectual standards of the people.

However, some Vietnamese museum at present could not grasp what is expected by the visitors, as a result, supplied information is not coincided with their expectations because limited such as:

Perform Science Education

In theoretically, museums must perform science education through disseminating science knowledge, intellectual information, which is studied, discovered and summarized from original exhibit collections to bring attractive and useful knowledge for the public. In the fact, however, some of museums only focused on merely education activities, almost given by upon dry, unattractive lectures. We are still not able to organize tours, education activities with specific targets, really excited visitors.

On the other hand, forms of culture service in museum are very poor, could not catch up with social objective demands.

In addition to, the supplying of social sciences and natural sciences knowledge is without synchronous and partially because the little amount is about natural sciences and technical.

Upon awareness of education role of the Vietnamese museum and its limitations, Prime Minister promulgated Decision No. 156/2005/QD–TTg dated June 23rd, 2005 about approval of overall planning for museum system so as to strengthen and develop museum system, serve for demands of science research, learning, teaching and dissemination of knowledge of history, culture, science and desires to enjoy culture of the public, as contribution into socio-economy development.

Introduction of Some Science Education Activities at Some Vietnamese Museums

"Imaginary Science Museum" within Vietnamese Museum of Ethnology

This online museum is called Kiosk Try Science, its contents were being built and supplied via web cams located at science centres all around the world, creating good conditions for children to discover the world of science via interactive pictures, multi-medium adventures, lively scene visits and taken into the Vietnamese museum of ethnology for exploitation.

From Kiosk Try Science, visitors shall make a journey through four missions: Archaeology mission, Space Exploration mission, Dolphin mission and Challenge mission.

Archaeology Mission

Acting as a biology archaeologist is searching for dinosaur fossils, you will have to discover and identify fossils before they are swept away by waters. You can find out excavation process of antiques by performing an excavation by yourself and uses fossils for reference to compare and identify fossils you have found.

Space Exploration Mission and Rescue International Space Station (ISS)

International Space Station is having trouble, you must try to prevent pressure reduction within spaceship chamber! In this mission, you will find out what should you do to become an astronaut is working on International Space Station when you transfer goods chamber from space shuttle to ISS, or explore living chamber of astronauts and get out of the ISS in case of emergency.

Underwater Wildlife–Explore with Dolphin

In this mission, you should help your friend, Dolphin to find prays and avoid dangerous flesh–eating animals as well as looking for flock of Dolphin. During mission performance, you will have the chance to find out lives of underwater animals, communication or position of whales by sound waves in the night and the dark water.

Figure 10.4: Dolphin Mission, Shockwave Game

Challenge: Race–track Conquer

You will participate into very difficult race–track with challenges so as to bring you knowledge of science and physics. You will compete with other competitors are accessing into online scientific museum from different places. That is why you are very hurry; otherwise competitors from other kiosks shall defeat you! In order to finish the race–track, you must overcome 5 challenges. These challenges shall teach you and inspect your knowledge of oblique movements, equilibrium theory, optical, reaction period and muscle.

Science Education of Geology in the Geological Museum

It was built in 1914, The Geological Museum (GM) is a museum with limited specialty in natural science, it plays an important part in studying, collecting, maintaining, keeping as archives, showing and propagating the knowledge about Vietnamese geology and minerals, and their historical development, as well.

Figure 10.5: The Geological Museum

In showing work, annual displaying system of GM is shown in building with 3 floors, with total area is about 1200 square kilometers and a part of atmosphere outside in front of the Museum. The displaying is arranged following by themes and contents:

Outside: to show some of patterns of stones, fossils, ores with huge size to create the characteristic of geology landscape of Museum.

The *first floor* is the theme of Vietnamese geological history and our planet with 5 contents followed:

- The Earth and Sun systems.
- Geological development of territory of Vietnam (include: thach quyen and biosphere).
- The process of basic geology (include: universe, dia dong luc, magma, tram tich, bien chat, phong hoa).
- The potential of Vietnamese minerals.
- Geology of Vietnam with Geology of International.

The *second floor* is the theme of Vietnamese geology and minerals, with 4 following contents:

- Regional geology (include: The North-Eastern, the North-western, the Central-Northern, the Highland, the Central-southern, and the Southern).
- The Minerals of Vietnam (include: metals, without metals, precious stones and a half of precious stones, fuels, radio-active, rare-soil, mineral water, hot water, and the source of glancing at heat).

Figure 10.6: Exhibits Showroom at 1st

Figure 10.7: Exhibits Showroom at 2nd

- Geology and Mineral of the East Sea and continent of Vietnam.
- Displaying all kinds of maps of Geology and Minerals. This floor is where displaying, sightseeing and studying, as well.

The *third floor* is place that collects some special subjects (like this: old- living things, mineral, stones, ores, and some kinds of product of geology). This floor is the place that shows, sightseeing and studying, as well.

Figure 10.8: Exhibits Showroom at 3rd

Science Education of Oceanography in Nha Trang

Oceanography Museum is built in 1927 at Nha Trang city–Khanh Hoa province–Vietnam. It is a unit that has long and organic relationship with Vietnamese Oceanography institution.

Oceanography Museum is a population continually, includes: tanks in which to bring up the sea creature to serve studying, sightseeing and educational community, as well and it is a system of building of keeping as archives the pattern things of the biggest sea creature in Vietnam.

With scale, to show the pattern of sea creatures with scientific combine with introducing about origin, their characteristics and special natures that is supplied by the result of studying of Oceanography institutional and other works of scientific studying from domestic and oversea have implemented successful the function of educational science of community in aspect of scientific of sea creature.

With one showing area that has universalized education about sea and sea-creature to community, such as: Sa Ban area, picture and models of sea ecological: Sa Ban: "Geography of continent of Vietnam": displaying the deep of sea bottom. The pattern: "sight of row of coral": to introduce the various of ecology in a row; and maps, pictures which introduce the source of benefit and propagate to protect the environment of the sea.

- The area of tanks with bringing up the sea creature inside and out side: to introduce about the various and diversifying of the sea creature, introducing about their origin and organic relationship helping each others between the sea creatures.
- The Area of diversifying ecology of Museum: to display 20.000 pattern things from tour of discovering in the East sea and some of neighboring seas with the aim to protect of the various of ecological of the sea creature. The pattern things are arranged to their development.
- The Area of displaying the big and small pattern is where showing the pattern of Shale's bone that is discovered and others were found in the area of Vietnam and neighboring.

The Future Natural Museum of Vietnam

According to the general plan of Vietnamese natural museum system by the year 2020, Vietnamese natural museum is a museum of national and arranged the first in the capable of the system of natural museum in Vietnam. With functions of studying, collecting, keeping as archives, displaying, propagating knowledge, introducing and advertising with the most enough, entire about the value of our natural country, ensuring the functional is a nationality natural museum, with has enough the ability of guidance, helping professional career for other museums, and also combining with high effective of actions in whole systems of natural museum of Viet Nam.

The natural museum of Vietnam includes museum and the garden of the national pattern things, the areas outside of displaying wall build on about 10 (hectare) at Me Tri commune–Tu Liem province–Ha Noi, in which 3 (hectare) built for the areas of inside displaying, which includes: universe, the earth and life, natural resources and the Vietnamese people, protection of recourses and the development of technology and science.

Now, the Museum is collecting the ideas of scientists to describe the displaying models about the value of natural resources of Vietnam.

Proposals for Cooperation

To strengthen more important role of science centres and museums in current context, we would like to propose some ideas on collaboration amongst member countries of NAM S&T Centre, as follows:

- Collaborative research in building a new science and technology museum in Vietnam.

- Sharing and learning experience on management and operation of science museums in the member countries of NAM S&T Centre.
- Training personnel for the science museums, to help them getting more experience and knowledge so that they could bring into play more role of scientific education through renovation of education activities in these museums.
- Strengthening the network of science museums in member countries of NAM S&T Centre, exchanging experts and exhibits amongst museums, etc.

References

The general plan of Vietnamese museum system until 2020 approved by the Prime Minister on 23/6/2005.

The introduction of Oceanography in Nha Trang.

The introduction of Geology in the Geological Museum.

The introduction of Vietnamese museum of ethnology.

Chapter 11

The Scientific Bases of Weather Modification: Popularising Applied Science through Specialised Centres

Mladjen Ćurić
Institute of Meteorology, University of Belgrade, Serbia
E-mail: curic@ff.bg.ac.yu

ABSTRACT

Man's aspiration to modify the weather is deeply rooted in antiquity. The stories of the Great Flood in Genesis and the plague of flood and storm in Exodus represent the absolute standard of the God's "abilities" to modify the weather. All subsequent weather modification efforts have been unimpressive in comparison with this standard. Despite of the impotence, the aspiration of ordinary people to control the weather emerged over and over in folklore and other proverbs. So, in order to bring the needed rain, primitive man used dances, various rituals and prayers. Praying for rain is still practiced in civilized as well as primitive societies. The current situation in weather modification is unique in that interest is based on new science discoveries, on new capabilities, and on careful and objective analysis of potentialities and limitations of any type of modification. Scientific weather modification had its beginnings in 1946. From that time the commercial weather modification projects are based on the achievement in the field of cloud physics. Agricultural crops protection based on the scientific-based weather modification methodology carries out in Serbia for more than four decades. This is one of the largest areas in the world. The operational weather modification project is carried out by the Meteorological Service. All the protected area is organized in so called polygons. There are twelve polygons. On each polygon there are good buildings with radar and other infrastructures. The programme of work is designed to operate 24 hours a day. Among the primary programme of the work, each polygon works as a science centre. They are the vital entity helping the people at large, school children and adults, in understanding and appreciating the scientific methodologies of weather modification. The objective of this paper is to review activity in Serbia that lead to the better exploitation of the weather modification infrastructures and ensuring sustainable development. This can be used as one model to enhance change by using the existing resources in each country.

Keywords: Weather modification centres, Crops protection, Rainfall enhancement, Applied science, Popularization.

1. Introduction

Weather modification is the effort of man to change naturally occurring weather for the benefit of someone. Through history and pre-history man has sought to modify weather by various means. Many primitive tribes have employed witch doctors or medicine men to bring clouds and rainfall during periods of drought and to drive away rain clouds during flooding episodes.

Numerous examples exist where modern man has shot cannons, rung bells, etc, in attempts to modify the weather. Superstitions have developed from time to time. Some were believed to have a scientific basis. So, Plutarch declared in the first century A.D. that: It is a matter of current observation that extraordinary rains pretty generally fall after great battles". In an attempt to rationalize cause and effect some people suggested that the noise of battle touched clouds of the rains (Changnon and Ivens, 1981). This theme was reemphasized after using the gunpowder up to the present days. Interest in weather modification based on this methodology started in the second half of the nineteenth century. At that time many patents were granted for making rain by explosions in clouds or for dissipating tornadoes by detonating an explosive in their centres. At the same time it was proposed to have rain production by the large fires. This proposal had a reasonable scientific basis. Namely, the large fires generate updrafts of air that, in a humid atmosphere, would support development of convective clouds leading to rain. A more extensive series of experiments of this type were carried out in Africa (Dessens, 1964).

The basis of scientific weather modification of clouds was established in 1933 when Tor Bergeron put forward a theory of rain based on the coexistence of ice crystals and super cooled droplets in clouds at the same temperature and on the rapid transfer of vapour from the droplets to the crystals (Bergeron, 1935). In 1946, it was discovered that the introduction of dry ice into a freezer containing cloud droplets cooled well below zero degree Celsius (what is called super cooled droplets) resulted in the formation of ice crystals. In 1947, a similar effect is demonstrated using silver iodide (AgI) particles. These experiments, which were carried out under the general direction of Nobel Laureate Irving Langmuir, launched us into the modern era of the science of weather modification.

It was well known at that time that under ordinary conditions, the formation (or nucleation) of ice crystals required the presence of a foreign substance called a nucleus, which would promote their formation. The basis of cloud physics is formation of precipitation particles (drops or ice crystals) on the particles of aerosols from molecules of water vapour and their growth by the mutual interactions. These interactions depend on the type of clouds. In warm clouds (a warm cloud is one that have wholly temperature above the 0°C the larger drops may increase in size by coalescing with smaller droplets. In cold clouds, (a cold cloud is one that have temperature below the 0°C at least in its upper levels) composed of a mixture of super cooled droplets and ice particles, the ice crystals may grow at the expense of the droplets.

On the base of these facts two methods of the artificially triggering of precipitation and dispersing of fog may be used:

1. The introduction into a cloud of large hygroscopic particles or liquid water droplets in order to initiate the coalescence mechanism of rain production;
2. Triggering of the Bergeron-Findeisen mechanism by producing ice crystals in a cold cloud by introducing artificial ice nuclei or some other suitable material. The second methodology may be used also for the hail suppression.

It was thus hypothesized that the insertion of dry ice or silver iodide in a super cooled cloud would initiate the formation of ice crystals, which in turn, would grow by vapour deposition into ice crystals. Precipitation could be artificially initiated in such clouds (Ćurić *et al.*, 2008). The very expression "cloud seeding" owes its origin to the fact that pieces of solid material placed in a supersaturated solution to promote precipitation of the dissolved solute or in a super cooled solution to cause it to freeze are known to physical chemists as "seeds". The artificial particles introduced into super cooled cloud droplets to induce freezing is an example of such seeds. The term "cloud seeding" invented to describe attempts to modify clouds by artificial nuclei. It is sometimes used today in a more general sense, to mean the release of any material designed to modify cloud properties.

An understanding of the possible outcomes of cloud seeding experiments must be based on the knowledge of the many branches of science, such is Meteorology, Physics, Chemistry, Electronics, etc. Also, the people involved in weather modification can be educated to use many very complicated technical tools. This knowledge, in turn, is based upon the results of the theory, laboratory experiments and observational data obtained with very modern equipments. The purpose of this paper is to review some of the most important methodologies used in the weather modification on the acceptable level for science centres.

2. Why Cloud Seeding in the Weather Modification

In this review obviously we must first explain why nearly all serious attempts to modify the weather to date have involved the seeding of clouds. The conclusions must be achieved taking into account that the all earth's atmosphere works as a giant heat engine. It uses the energy mainly from the sun to drive the winds and lift the water vapour from the ocean surface to deposit it as a precipitation (rain or snow) on the earth's surface. Very little solar energy is absorbed directly by the atmosphere. The solar energy is absorbed at the earth's surface and then enters the lowest layers of the atmosphere as a sensible heat or as a latent heat associated with the evaporation of water.

The solar energy absorbed by the earth enters only the earth's radiation balance, offsetting the energy radiated away into space, either from the atmosphere or the surface itself. Only a small part of the bulk solar energy which is relayed to the atmosphere ever appears as a kinetic energy associated with the winds. The energy of the winds is derived principally from the differential heating of different parts of the earth's surface. Some solar energy is stored for a time as a latent heat of evaporation, to be released later when the vapour condenses into clouds. Local thunderstorms and tropical hurricanes are powered by the release of latent heat of evaporation.

The kinetic energy of the winds (energy of the air motion) is small compared to the total of energy reaching from the sun each day. However, it is still so vast that any attempt to change the weather by direct applications of energy artificially produced by men has no chance of success. For example, air conditioning of a coastal city by giant fans on a day contrary to the natural air circulation would be prohibitively expensive. On the contrary, extract energy from the wind by giant wind mill farms is possible without the slow down of the wind due to this energy extraction, or otherwise influence weather conditions. Obviously, the cheap weather modification methodology must involve the modulation of the release of energy already stored in the atmosphere or the modification of radiative energy transfer. A possible way to accomplish the later would be seeding of the clouds or clear sky to change the reflecting characteristics of parts of the earth's atmosphere or surface.

3. Organization of Weather Modification in Serbia

Some types of the weather modification projects in Serbia are carried out by the Meteorological Service from 1967. The whole area is divided in so called polygons (Figure 11.1). Each polygon has a radius of approximately 30 km. The programme is designed to operate 24 hours per each day. All clouds over the polygon are monitored by S band (10 cm wave length) radar located at the centre of the polygon (Figure 11.2). The radars are conventional, Doppler and polarized. Their technical characteristics are as follow: transmitting power 600 KW, pulse length 1/2S, wave length 10 cm, beam width 1.1°, antenna gain 42.3 dB and antenna reflector 6.1 m.

When a monitored cloud meets specific criteria based on the radar reflectivity, the level of the significant temperatures and the ration of the depth of cold and warm part of cloud, the clouds are seeded with glaciogenic seeding agents (silver iodide particles). The seeding agents are delivered into target area by rockets fired from the ground based network of firing stations separated by about 5 km.

The firing stations are manned by the trained volunteer observers. The responsibilities of the observer, among the firing the rockets, is to observe and reports days with lightning, thunder, hail and others precipitation type (ranged in nine classes). Such dense network of the firing stations is a special mesometeorological network.

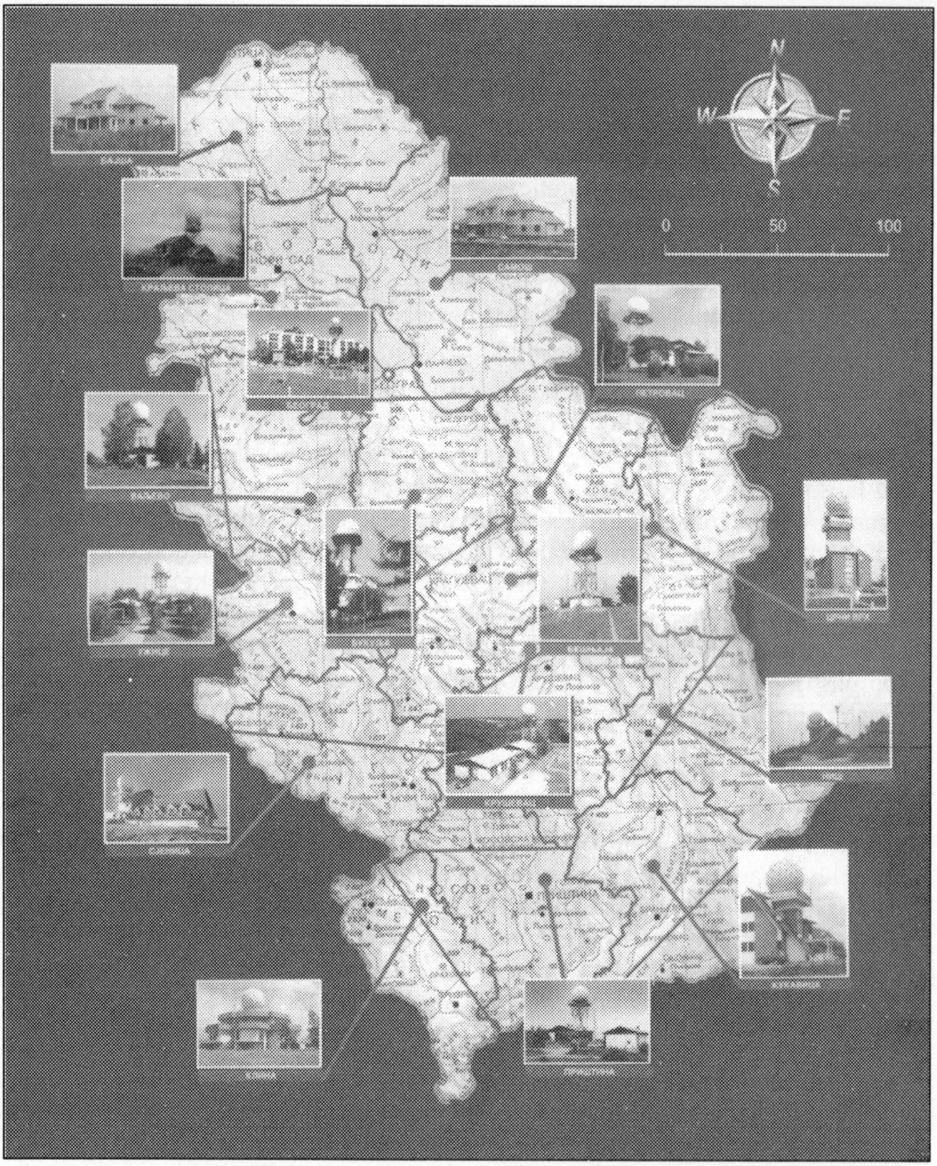

Figure 11.1: The Network of Radar Centres in Serbia

The rockets fires from the ground based rocket launcher (Figure 11.3). Each launch sites needs at least one arsenal for storing the rockets. It is notable that personnel requirements are very intensive as there must be trained staff whenever appropriate clouds are forecast.

The rockets are designed to release a plume of the nucleating material along a trajectory in the cloud (Figure 11.4). Launch crews receive details about the necessary elevation and azimuth angles from the radar operators, or the targeting may be

Figure 11.2: Meteorological Radar on Bukulja Polygon

Figure 11.3: The Operator on the Ground Based Rocket Launcher

Figure 11.4: The Weather Modification Rockets

automated by telemetry of aiming and firing instructions to the launcher. The greatest advantage of the rocket system is its ability to reach any region of the targeting cloud even it a dangerous thunder cloud in a very short time. Rockets will reach the seeding zone in several seconds and the entire trajectory is completed in less than one minute.

4. The Radar Centres as a Science Centres for Weather Modification

Science centres are very important for the rapid advancements of science and technology among the ordinary people. The children and adults can learn about the natural laws, technological development, etc. One of the very important parts of our environment is the atmosphere. The understanding of its natural behaviour and the possibilities for the modification are crucial for redirecting the people away from superstitions to the scientific scene. The weather modification centres are equipped with many contemporary tools (radar, lidar, instrumented airplane, etc). So, they are the places where one can discover, explore and test ideas about the nature of atmosphere.

Participants learn because clouds contain stored energy, they constitute a promising place for attempts at weather modification. The fact that clouds yield practically all of the atmospheric water deposited on the surface of the earth makes their modification a matter of great practical importance. These considerations make it clear why nearly all deliberate weather modification attempts so far have been directed at clouds. Some methods for modifying weather are described in greater detailed below.

4.1 The Modification of Warm Clouds

In tropical and semi-tropical regions, in particular, as well as in extra-tropical regions clouds that are too shallow to extend into freezing levels are prevalent. During drought periods, if any clouds form at all, they are generally the warm and non-precipitating clouds. The motivation is therefore strong to develop methods for extracting rainfall from non-ice- phase clouds.

The process of precipitation formation in warm clouds begins with the nucleation of cloud droplets on hygroscopic aerosol particles. These are airborne dust particles that are normally quite small (less than one micrometer in diameter) and have a natural affinity for water vapour (salts such as ammonium sulphate, sodium chloride and natrium chloride particles). On these particles the droplets grow by vapour deposition (Figure 11.5a). But, vapour deposition growth process is to slow for the formation of warm rain.

The growth of precipitation- size droplets (drop of diameter of 1 mm) in some warm clouds may be restricted by the lack of sufficient numbers of large condensation nuclei. It seems feasible therefore those certain warm clouds might be induced to rain by introducing into the cloud large hygroscopic particles.

Warm rain formation is dominated by collision and coalescence among droplets (Figure 11.5b). A more direct method of modification of droplet spectrum of natural clouds is to introduce water droplets directly into the cloud. Theoretical considerations show that the best level at which to introduce the droplets is cloud base, since in this case they may be carried on updraft until they have grown to sufficient size to fall out of the cloud as precipitation. The optimum droplet size for seeding depends on the speed of the updraft and the dept of the cloud. For example, a droplet of initial radius of 30 micrometers, carried up from the base of a cloud where the temperature is 20°C in a uniform updraft of 3 m/s, will return with a radius of 1.9 mm. The corresponding mass multiplication is 250 000. Experiments indicate that under certain conditions the seeding of warm clouds with hygroscopic nuclei or water droplets produce rain.

4.2 The Modification of Cold Clouds

By far the greatest effort to increase rainfall from natural clouds has been directed at triggering the Wegener-Bergeron-Findeisen mechanism of rain formation in cold clouds (Bergeron, 1935). These three scientists showed that ice crystals once formed in a super cooled cloud could grow very rapidly by deposition of water vapour onto them at the expense of super cooled cloud droplets. This is due to the fact that the saturation vapour pressure with respect to ice is lower than the saturation vapour pressure with respect to water at temperatures colder than 0°C. Thus, an ice crystal

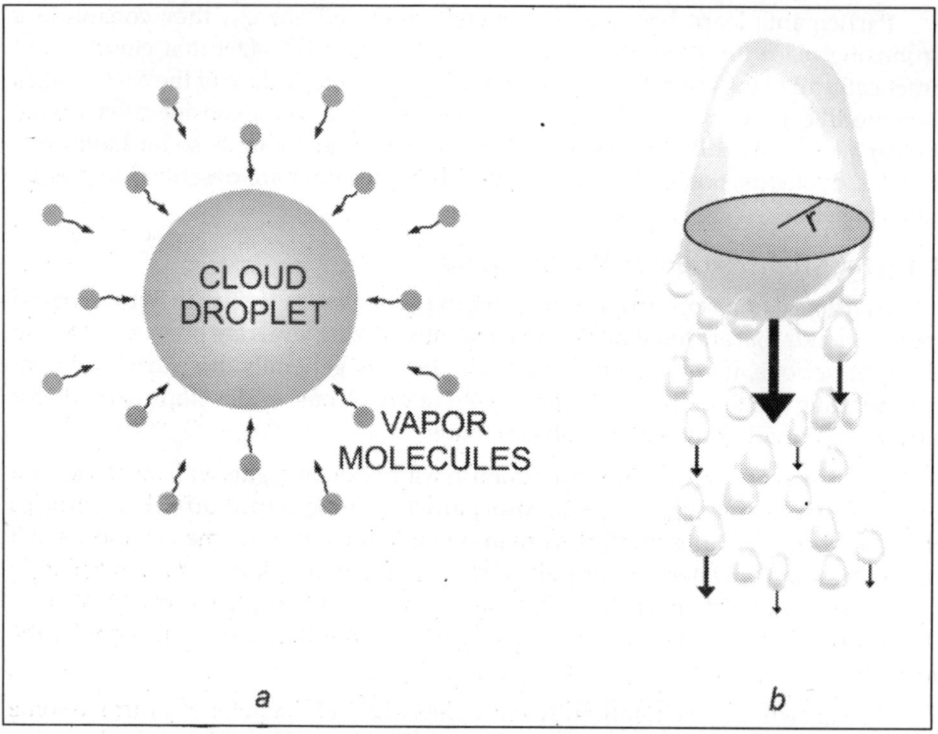

Figure 11.5: (a) Droplet Growing by Deposition of Water Molecules (b) Growth of a Drop by Colliding and Coalescing with Smaller, Slower-Settling, Cloud Droplets

nucleated in a cloud that is saturated finds itself in an environment which is supersaturated with respect to ice, and can thereby grow rapidly by deposition of vapour (Figure 11.6). As vapour is deposited on the growing ice crystals the vapour in the cloud is depleted, and the cloud vapour pressure lowers to below water

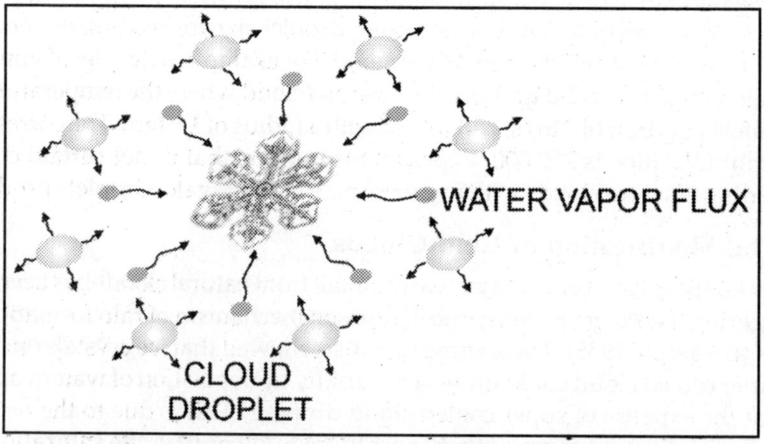

Figure 11.6: Schematic Illustration of the Wegener-Bergeron–Findeisen Process of the Ice Crystal Growth

saturation. Thus, cloud droplets evaporate providing a reservoir of water vapour for growing ice crystals. Therefore, the ice crystals grow at the expense of the cloud droplets.

The concentration of natural ice nuclei in the atmosphere is very small, and on some cases is probably less than required for the efficient initiation of the Bergeron-Findeisen process. Under these conditions it should be possible to induce a cloud to rain by seeding it with artificial ice nuclei, or some other material that cause ice to appear. This is, in fact, the scientific bases of most of the rainmaking experiments that have been carried out in the past sixty years. The materials suitable for seeding cold clouds are silver iodide, copper iodide, dry ice (solid carbon dioxide), propane, etc.

4.3 Cloud Seeding for Hail Suppression

Thousands of years man sought methods for suppressing hail. This has ranged from ringing bells, firing cannons to the modern era of cloud seeding. The motivation for suppressing hail is strong, since hail can devastate a wheat, orchard, or grape cop. When hail storm occurs over a modern urban area the huge property damage can result and the humans and livestock can be severely injured or even killed. Scientific programmes aimed at developing a hail suppression effort using cloud seeding strategies began in the 1950's in the Russia, Serbia, etc. (Ćurić, 1990; Radinovic, 1972).

In order to understand how hail forms, one must first understand the varying behaviour of thunderstorms which are the factories where hail stones are produced. A thunderstorm produces the liquid water content upon which hailstones grow. The updrafts then suspend the embryo hailstones and determine how long a growing hailstone will remain in a water rich environment.

The methodologies of suppressing of hail vary considerably depending on the conceptual model one considers the most appropriate for the hailstorms in a different region of the world. Differences in conceptual hailstorm models results from: differing environmental conditions and from different perceptions of hailstorm structure. Perceptions of hailstorm structure, in turn, vary from one scientist to another and are dependent on the hailstorm observing systems and numerical models available to a researcher. For example, one's perception of a hailstorm structure changes as one moves from observing a hailstorm with a radar providing only reflectivity values, to observing a hailstorm with multiple Doppler radar and a multiparameter radar (polarized radar) capable of also identifying ice versus liquid precipitation elements (Bojkov, 2003).

One hail suppression technique is based on the direct injection of silver iodide seeding material in the region of high radar reflectivity and presumably large amounts of super cooled water. The technique involves the use of several types of hail rockets which carry seeding material to heights as much as 8 km where material is explosively dispersed over the target region (Figure 11.8).

This technique is used in Serbia. Another proposed approach to hail reduction is to seedthe lower part of the clouds with salt particles, or some other hygroscopic material, and thereby initiate a warm rain process in the lower levels of the cloud.

Figure 11.7: Participants at the Radar Centre Discuss the Radar Characteristics of One Hailstorm

Figure 11.8: Launching of Hail Rocket with the Ground Based Rocket Launcher

This concept is known as the trajectory lowering technique (Ćurić and Vukovic, 2005; Vukovic and , 2005).

5. Conclusions

The special attention is devoted to the description of the scientific bases of weather modification. Many weather events are responsible for more outdoor deaths and property loss than other violent forces of the nature. Due to this reason it is very important to educate the children and adults about the natural lows of such processes in the atmosphere and about the possibility for the modification. The understanding of its natural behavior and the possibilities for the modification are crucial for redirecting the people from a way of the superstitions to the scientific scene.

Science centres are very important for the rapid advancements of science and technology among the ordinary people. The children and adults can learn about the natural laws, technological development, etc. at different types of the institutions. Here is described how weather modification centres in Serbia serve as science centres in the development of the scientific methodologies in this very important part of our environment.

In our analysis we have paid attention to the description of the methodologies for the precipitation enhancement in the warm and cold clouds and hail suppression. There are two major reasons for this work. It is envisaged that this scientific programme would greatly benefit the science educators and experts involved with science popularization, and for launching of the new types of science centres.

Acknowledgements

The authors would like gratefully to acknowledge to the Hydrometeorological Service of Serbia for providing necessary data and Mr. Dragomir Bulatovic for his help in preparing the figures.

References

Bergeron, T., 1935: On the physics of clouds and precipitation. *Proc. 5th Assembly UGGI, Lisbon*, 2, pp.156-161.

Bojkov, R., 2003: Brief review of current status of hail suppression. *WMO, Report Series No. 41, Report No.* 6, pp.49-55.

Changnon, S.A. and Ivens, J.L., 1981: History repeated: The forgotten hail cannons of Europe. *Bull. Amer. Meteor. Soc.*, 62, pp.368-375

Ćurić, M.,1990: One evidence of the hail suppression efficiency derived from radar measurements. *J. Wea. Mod.*,19, pp.79-81.

Ćurić, M., and Vukovic, Z., 2005: The influence of lightning on precipitation revealed by mesoscale numerical model. *IAMAS Abstracts, Beijing, China*, G2.

Ćurić, M., Janc, D.and Vuckovic, V., 2008: Precipitation change from a Cumulonimbus cloud downwind of a seeded target area. *J. Geophys. Res.*, 113, D11215, doi:10.1029/2007JD009483.

Dessens, H., 1964: Physics of precipitation. *Geophys. Monogr,Washington, D.C.*, 5, pp.396-398.

Radinovic, D., 1972: Hail Control–Operative and scientific research project. *U.S. Department of Commerce and National Science Foundation, Washington*, pp. 124.

Vukovic, Z., and Ćurić, M., 2005: The influence of the acoustic-electric coalescence on phase and mass transfer of supercooled drops. *Pure Appl. Geophys*,162, pp.1-25.

Part III

Rationale for the Establishment of Science Centres

Chapter 12

Potential Contribution of Science Centres to Performance in Schools

Thato Bolila[1] and Mpho Sekhosana-Nyenye[2]

[1]Department of Science and Technology, Lesotho

E-mail: missbolila@yahoo.com

[2]Ministry of Education and Training, Lesotho

E-mail: mphojs@yahoo.com

ABSTRACT

This paper starts by looking at the prevailing situation in Lesotho, in the field of Science and Technology and initiatives taken by the government of Lesotho, in trying to curb the situation, and what the government is still doing to redress the situation. It then looks into the rationale to establish a science centre, the strategic approaches and the expected impact of the science centre. Science and technology are regarded as engines of the socio-economic development of any nation and medium and long-term economic returns of any country can be realized through investing in science and technology. In Lesotho and many other African countries, science and technology are characterized by the high failure rate which leads to low enrolment at tertiary level. Although the government of Lesotho has taken some initiatives to improve the performance in science the problem still persists. Students' performance in science at secondary level is still not satisfactory. It is expected that establishment of a science centre would improve performance in science; hence increasing enrolment in science and technology at tertiary level. It is also hoped that this could be achieved through the professional development of teachers, individual and group counselling of students, career mentoring of students, making science equipment available to disadvantaged schools and holding science-enriched programmes for students who have potential in science subjects.

Keywords: *Science Centre, Performance, Career guidance, Counselling, Teacher professional development.*

Introduction

Global economic history has inevitably shown that, science and technology are the engine of socio-economic development of any nation (Tan and Subramanian, 2003). Furthermore, progress in science and technology education eventually influences our daily lives by becoming technologically dependent. It is through investment in science and technology that the medium and long-term economic returns can be realized.

The prevailing situation in the field of science and technology in Lesotho is characterized by the high failure rate at secondary level and consequently, low student enrolment in science subjects at tertiary level leading to a low human resource base in science and technology labour force. Science subjects are perceived as difficult and hence the majority of students at tertiary level resort to other subjects. Establishment of a science centre in Lesotho is a vital mechanism for shifting the student's paradigm as far as science subjects are concerned.

Nevertheless, in the past years, there have been successful initiatives to popularize science education. An example is the Lesotho Science Pre-entry Course (LESPEC). The aim of LESPEC was to recruit students in science subjects, who were otherwise likely to perform badly at secondary level, and prepare them for entry into tertiary institutions. Furthermore, it also helped those students who could do well in the sciences, but were not given the opportunity to study the three science subjects (Physics, Chemistry and Biology) in their respective schools. Likewise, the introduction of Accelerated Mathematics and Science Teachers' Induction Program (AMSTIP) helped the teachers to acquire extra skills in the teaching of science subjects.

The Government of Lesotho took further efforts to redress the situation. Its commitment to leverage science and technology for socio-economic development is well articulated in the Vision 2020, National Science and Technology Policy (2006), the Education Sector Strategic Plan 2005 and Education-for-all policy, to mention but a few.

It is on this note that the Department of Science and Technology is embarking on establishing a national science centre through its programme of promoting Public Understanding of Science Engineering and Technology (PUSET).

Research studies carried out to assess the impact of science centres in different countries has shown that science centres are able to;

- Provide a range of activities which have some educational content.
- Provide access to science as they attract a wider audience.
- Create dialogue by breaking down existing barriers between scientists and the public.
- Raise awareness and interest
- Create memorable learning experiences
- Provide long term learning

(http://www.ecsite-uk.net/resources/reports/wellcome-trust-impact-assessment.pdf; http://www.ecsite-uk.net/reports/downloads/impact-of-science-discovery-centres-review-of-worldwide-studies.pdf)

Rationale Towards Establishment of a Science Centre

The current statistics in the performance and enrolment in the Mathematics, science and technology clearly indicate that there is a need for other interventions? To be put in place to promote science and technology studies through informally educating (edutainment) and stimulating subjects in a participative manner. Moreover, limitation to career guidance confine students in exploring other avenues available in science and technology education. Consequently fully flushed career counseling will help students to discover their full potential and preferences prior to enrollment.

In addition, enrichment programmes will expose students to science and technology activities, which will include hands-on laboratory work, interact with scientists and researchers on mini projects, and also work on solving science and mathematics quizzes. As a result, this is envisaged to improve the pass rate and enrollment in science and technology education by students.

The enrichment programme will also acquaint teachers with new materials and innovations, which will provide them with more information and resources needed to enable them to deliver their services most effectively.

The present trends in the performance of sciences in secondary level indicate that pure sciences are not well performed, and even the candidates who pass them do not opt for them at tertiary level, while pure science are important in skills development of the students (Table 12.1)

Table 12.1: Trends of Science Performance at High School Level (2002-2005)

Subject	*Year*	*Number of Students and their Corresponding Symbols*							
		A	*B*	*C*	*D*	*E*	*F*	*Total*	*% Pass*
Biology	2002	6	107	430	500	677	2650	4370	39.4
	2003	18	158	424	521	603	2564	4288	43.7
	2004	11	197	435	511	613	2262	4029	44
	2005	55	316	523	484	584	1933	3895	50
Science	2002	24	111	231	439	492	587	1884	69
(Bio/	2003	20	115	242	506	526	483	1893	74.4
Chemistry)	2004	19	153	291	555	694	669	2381	72
	2005	49	161	264	494	576	902	2446	63
Science	2002	98	292	468	705	746	980	3289	70
(Physics/	2003	139	294	428	656	624	924	2996	71.5
Chemistry)	2004	121	300	437	692	717	1075	3342	68
	2005	166	380	453	616	721	1150	3486	67
Combined	2002	88	235	302	239	194	160	1218	87
Science	2003	96	303	302	228	182	138	1249	88.9
	2004	72	205	337	258	171	134	1177	89
	2005	132	297	334	247	196	257	1463	82

Contd...

Table 12.1–Contd...

Subject	Year	Number of Students and their Corresponding Symbols							
		A	B	C	D	E	F	Total	% Pass
Add.	2002	55	104	131	192	218	226	924	76
Combined	2003	56	122	130	182	183	244	844	79.7
Science	2004	41	92	142	170	231	248	924	73
	2005	43	138	170	241	285	232	1209	73

Source: Ministry of Education and Training results statistics of 2005.

A,B,C,D,E and F; symbols, A > 85 per cent, B > 75 per cent, C > 65 per cent, D > 55 per cent, E < 50 per cent, F < 40 per cent.

Table 12.1 indicates the general trend on secondary results in science subjects from 2002 to 2005. It is clear that there has been a decline in students' enrollment rate in the sciences and a high rate of failure. Likewise, the pass rate in percentage may seem higher but the average pass for all the candidates was 65.4 per cent and only 8 per cent of these candidates were able to obtain credit (A, B or C). This will mean that only 8 per cent of students will have access to university education and the rest of the students will have to seek admission else where. Of the 8 per cent, only students doing Science and Biology will be the ones admitted to do science related courses.

Below is percentage performance represented in charts in different sciences:

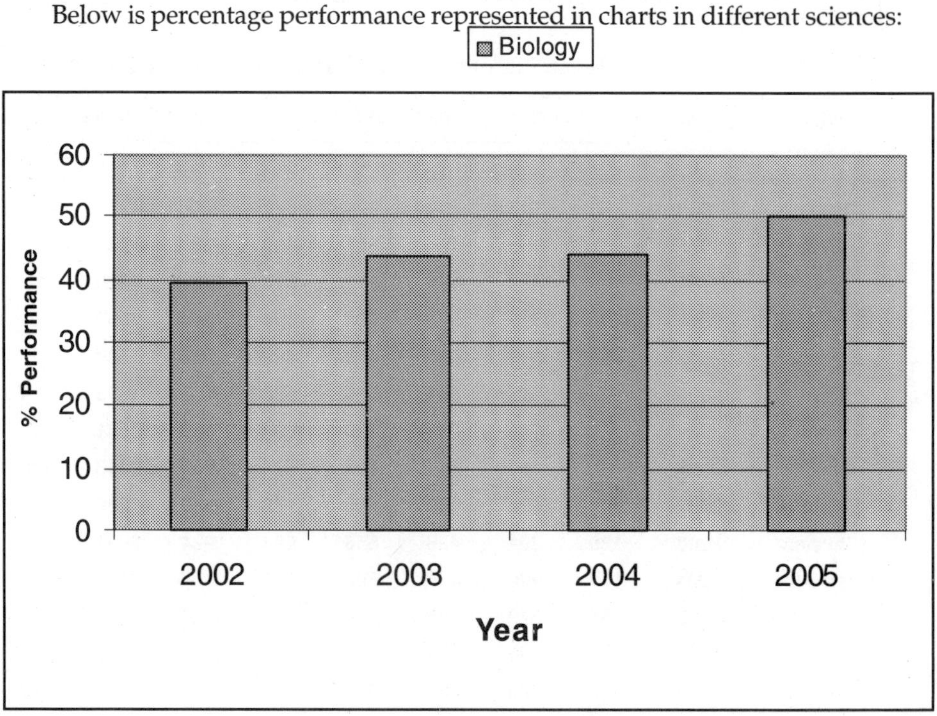

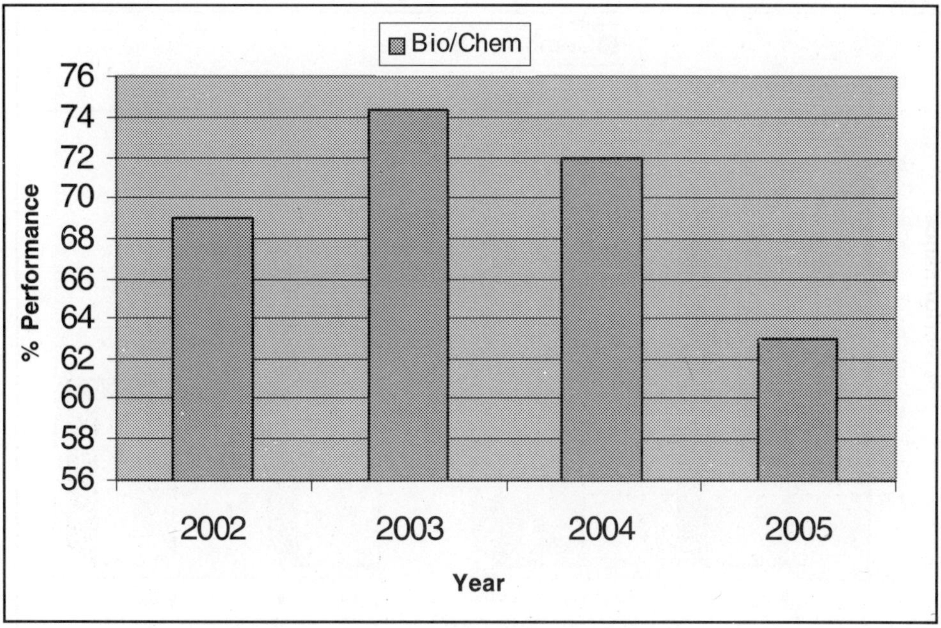
Bio/Chem
76
74
72
70
68
66
64
62
60
58
56
% Performance
2002
2003
2004
2005
Year

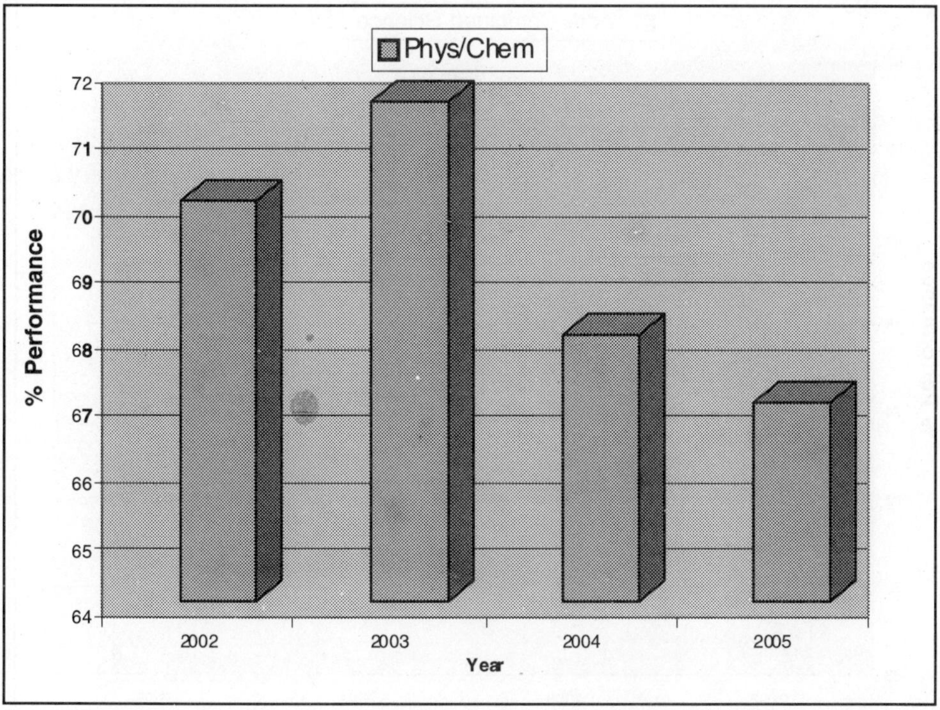
Phys/Chem
72
71
70
69
68
67
66
65
64
% Performance
2002
2003
2004
2005
Year

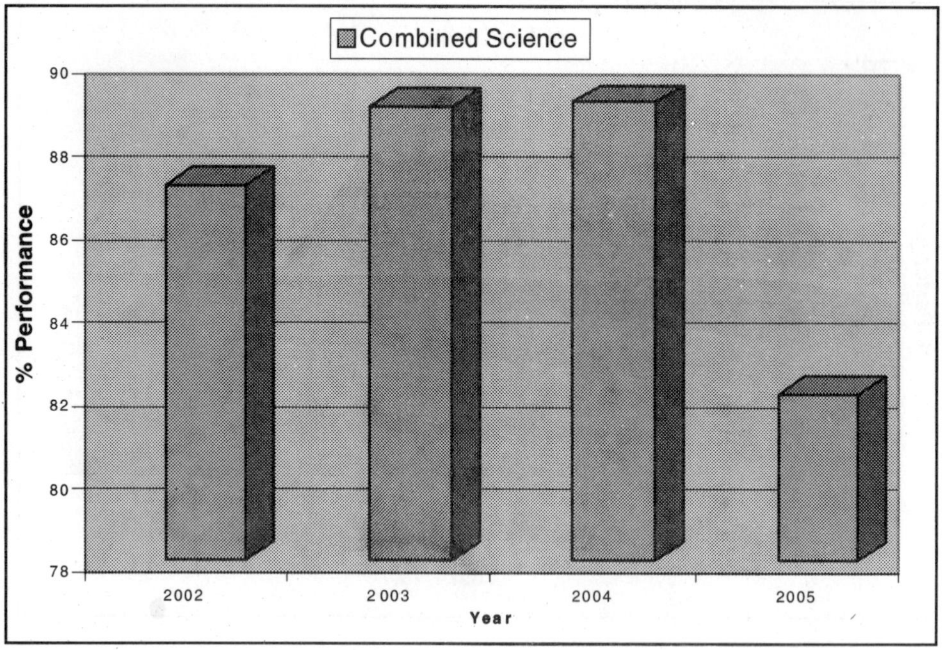
Combined Science
90
88
86
84
82
80
78
% Performance
2002
2003
2004
2005
Year

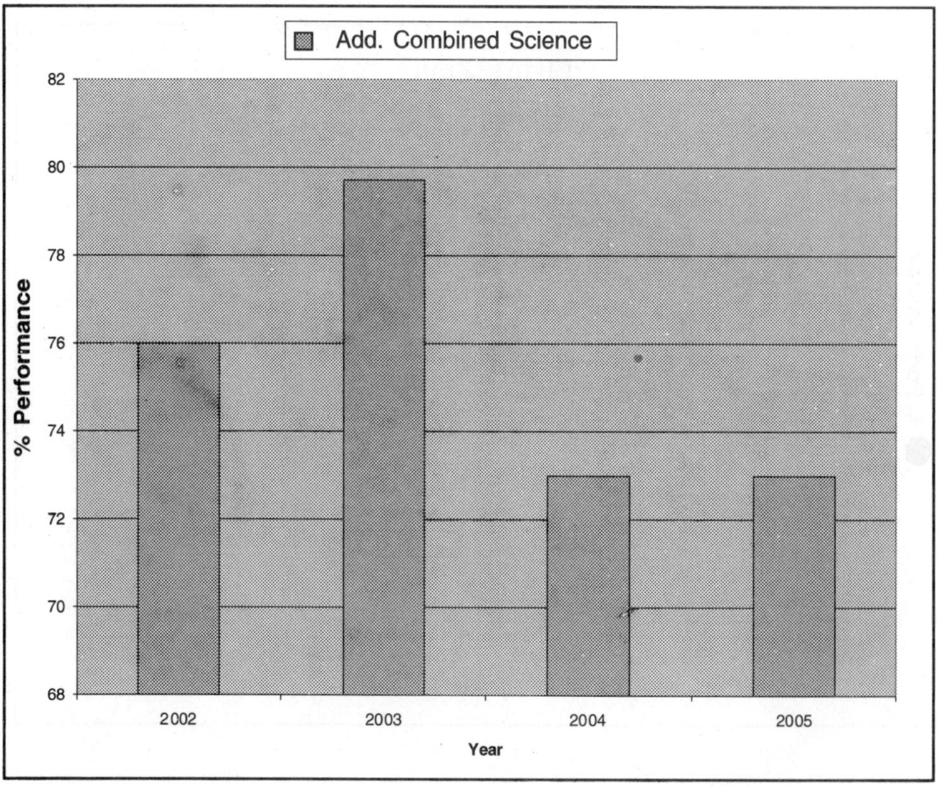
Add. Combined Science
82
80
78
76
74
72
70
68
% Performance
2002
2003
2004
2005
Year

Strategic Approach to Enhance Change through Science Centres

As indicated through statistical analysis, performance in science and technology education is characterized by a high failure rate which indicates a need to curb the situation. The envisaged science centre will serve the following functions:

Promoting Public Understanding of Science and Technology (PUSET)

Through the production of educational materials and newsletters for information dissemination, public understanding of science and technology will be improved. Again, a national science and technology week will be celebrated at the centre including the hosting of a variety of science and technology activities, such as Physics day, Astronomy months and many others. Science centres has been found to have a positive impact on visitors. In one of the studies carried out by Allen in 2002, it was found that visitors were able to engage in conceptual conversations at over one third of the exhibitions (http://www.ecsite-uk.net/reports/downloads/impact-of-science-discovery-centres-review-of-worldwide-studies.pdf).

Career Guidance Counselling Centre

(i) *Individual and group Career counseling* plays a vital role in helping students to realize the careers available in science and technology and employment opportunities available for scientists. As an extension service, group counselling will be offered to visiting schools, from all districts in Lesotho, at the centre.

(ii) *Career mentoring programme* will be a self-managed and computerized programme aimed at helping students to know their interest through using computer-based tests to discover career opportunities on the database.

(iii) *Psychometric Assessment* will also be necessary to evaluate ability, interest, values and personality of individual student, and assist students on their preferences and strengths.

Provision of Resources for Disadvantaged Schools

Most of secondary schools in Lesotho do not have laboratories and even those that have laboratories, are not fully flashed with up to date equipment. The science centre through its outreach programme will provide mobile exhibits in different districts of Lesotho. This means that some of the exhibits in the centre will be movable and available to students in all districts.

Furthermore, the centre will provide science shows, which are interactive science experiments and quizzes. It is also hoped that, beside the mobile exhibits, the centre will have laboratory equipment and space which will be used by visiting schools to perform experiments. This will assist schools that have no laboratories. At present there are about 310 secondary and high schools and about two thirds of these do not have laboratories. This hinders effective learning and encourages teacher centred teaching approaching. The availability of mobile exhibitions is hoped to assist these schools.

Host Science Enrichment Programmes

As indicated, there is a low student turnover in science and technology subjects. These programmes will be offered on weekends as sessions for students with potential in science subjects who seem to have a low passing rate. Demonstrations and workshops will be carried out, which are in line with the syllabus, for these students. Again, tutorials will be provided to assist these students to understand different concepts.

Secondly teacher's forum aimed at assisting science teachers will be part of enrichment programme. These programmes are expected to be held monthly to give teachers an opportunity to meet, share ideas and discuss problems they encounter in the classrooms. Again the centre will be used for in-service training of science teachers which at present is a mandate of science inspectors.

The in-service training will be held for all the science teachers in the country during holidays or on a weekly basis for teachers who need assistance. This will depend on teachers' demands. Teacher educators and other relevant bodies will be invited to assist these teachers.

Lesotho like some other countries do not have a pool of qualified science teachers, even the ones that are available are always looking for greener pastures. Institutionalized in-service training will assist to equip unqualified teachers with contend and teaching methods required in classrooms.

Science enriched programmes had been found to improve students and teachers' understanding of science concepts. In a study carried out by the Wellcome Trust, in 2007 to assess the impact of science centres, it was concluded that:

"Science centres are able to offer a very different learning experience to that of the classroom. The majority of teachers who participated in the impact assessment spoke positively about the science centres, commenting that their trips: "helped to create a positive attitude towards science in my pupils"; "brought [science] to life"; "made science real"; "help[ed] the kids to engage"; and "helped children visualise the curriculum".

(http://www.wellcome.ac.uk/stellent/groups/corporatesite/@policy_communications/documents/web_document/WTD038282.pdf).

Workshop Facility

Learning takes place as learners manipulate and touch the hardware of technology and through self-discovery. Therefore, initially the centre will buy exhibits from existing manufacturers. Further, as the life cycle of the exhibits declines, learners will need other exhibits challenging or in line with the syllabus. Also as a result of playing with the exhibits, it is expected that they will need to be repaired; therefore a fully equipped workshop is a necessity for replacing worn-out exhibits and making new ones.

The Expected Impact

It is expected that, like in many countries such as South Africa, Singapore and others, the science centre will:

- Increase interest and initiative among the youth in science and technology careers.
- Enable students to make well informed decisions in science and technology careers through guidance and counseling.
- Improve understanding of science concepts through demonstration of application of science, engineering and technology in the real world.
- Lead to well supported and managed science education by provision of training to science teachers and outreach programmes to students.
- Provide better understanding of science and technology amongst youth and public at large through dissemination of information.
- Increase scientific and technological workforce overtime.
- Lead to a knowledge driven economy of Lesotho by the year 2020 as perceived by the national vision.
- Improve science and technology literacy in Basotho.

Conclusion

Establishment of Science Centre is a necessity as it will add value in enhancing the love for science among students in Lesotho. This initiative will also supplement other existing measures taken by the Ministry of Education and Training, the Ministry of Communications through Department of Science and Technology and LSMTA, to improve science performance in primary and secondary schools. It will also assist the Ministry of Education and Training to achieve the national Vision 2020 which is based on achieving a technologically well equipped Lesotho by 2020. Given the performance of students at different levels and their participation in science subjects as opposed to other subjects, there is a clear indication that other strategies to demystify science are a responsibility of the country. This project is feasible and will be an invaluable investment, as it will pay dividends in terms of capacity building in science and technology, as these are the driving pillars for the Lesotho's economic growth and its sustenance.

References

Government of Lesotho. Ministry of Education and Training, 2004-2005. Performance in Science Report. Department of Central Inspectorate. Maseru. Lesotho.

Government of Lesotho, Ministry of Finance and Planning, 2004. Lesotho Vision 2020. Morija Printing Works. Lesotho.

Government of Lesotho, 2006.Science and Technology Policy. Vumo Holding printers, Maseru Lesotho.

http://www.wellcome.ac.uk/stellent/groups/corporatesite/@policy_communications/documents/web_document/WTD038282.pdf

http://www.ecsite-uk.net/resources/reports/wellcome-trust-impact-assessment.pdf

http://www.ecsite-uk.net/reports/downloads/impact-of-science-discovery-centres-review-of-worldwide-studies.pdf

Ministry of Education and Training, 2005-2015. Education Sector Strategic Plan, BeePee Printers and Stationary.

Tan, L. W. H. and Subramanian, R., 2003. 'Science and technology centres as agents for promoting science culture in developing nations'. Int. J. Technology Management, Vol. 25, No. 5, pp.413-426.

Yearley, S., 1995. Misunderstanding The Public: The Public Understanding of Science Movement. *Science, technology and innovation*, December 1995, pp. 25-29.

Chapter 13

Status of the Establishment of Science Centres in Malawi

Rex Kanjedza
Department of Science and Technology, Lilongwe 3, Malawi
E-mail: rexkanjedza@yahoo.co.uk

ABSTRACT

Malawi, like the rest of the world, recognizes the major role Science Centres play in the development of science, technology and innovation. Science Centres are the information centres for raising public awareness of current research and stimulating an interest in science among people, especially the youth. This is usually done through galleries, displays, exhibits, lectures among others. The National Research Council of Malawi prepared a proposal to establish first science centre in Lilongwe, the capital city of Malawi in 2004. The project was aimed at creating and promoting the public understanding of science, engineering and technology; supporting, enriching and complementing the existing formal provision of science education through providing an informal approach to science teaching and learning; creating public awareness of science and technology; promoting science and technology culture; and promoting science communication and knowledge. The project proposal was then submitted to development partners since government of Malawi was resource- constrained. The need for the establishment of the science centres is pressing. Malawi's current vision is to transform from a predominately consuming and importing nation to a predominantly producing and exporting one. The establishment of science centres, therefore, is regarded as one of the tools that can contribute significantly to the development of science and technology, which is a vehicle for the realization of the vision. The situation on the ground reveals that the science and technology culture in Malawi is not quite developed. This is evidenced by the noticeable weak picture of public understanding of science, engineering and technology that often result into few people, especially the youths, being stimulated into making careers in science and technology. The proposed project was to benefit a cross-section of members of the public. These include young and old, male and female, school going and non-school going and persons with disabilities by promoting informal teaching and learning of science and technology. The science centre was to target beneficiaries/audience through visits that members of the public would be making to the centre and through the centre's outreach activities. The establishment of the centre was expected to see an increase in the number of students taking and doing well in science and technology related courses.

Keywords: *Malawi, Science centre, Technology, Innovation, Establishment.*

1. Introduction

The concept of establishing Science Centres in Malawi was conceived after Malawi attended the 3rd Conference of Southern Africa Association of Science and Technology Centres in November 2000, where participating countries were urged to consider setting up *'Science and Technology Centres'* as a national priority. In 2004 a proposal to establish the science discovery centre was developed by the National Research council of Malawi and was submitted to our development partners since the government of Malawi was resource-constrained. Unfortunately, the proposal has never been funded up to now. Currently, Malawi does not have Science Centres.

Malawi's current vision is to transform from a predominately consuming and importing nation to a producing and exporting one. Science and Technology is therefore seen as one of the tools that can be used to realize this vision through the promotion of the development of competitive value added services and utilization and development of both indigenous and exotic technologies. It is therefore believed that the establishment of Science Centres in the country can contribute significantly to the development of Science and Technology by enhancing public understanding of science, especially amongst the youths and stimulating them to pursue careers in science and technology.

1.1 Objectives for Establishing Science Centre

The main objective for establishing Science Centre is to promote the public understanding of science, engineering and technology in Malawi by supporting and complementing the existing (formal) provision of science education through providing an informal approach to science teaching and learning.

1.2 Proposed Functional Activities for the Science Centre

The centre's functional activities shall be to:

1. Carry out interactive science and technology exhibitions and demonstrations with scientific exhibits/models that illustrate scientific principles within a science centre;
2. Run interactive educational and promotional outreach/mobile science exhibitions and demonstrations;
3. Support schools (*i.e* teachers and students) in conducting curriculum and non-curriculum based hands-on science experiments and demonstrations thereby providing not only an opportunity for enhancing teacher and

students skills and experiences in conducting experiments but also providing an opportunity for students to learn scientific principles;

4. Mobilize and support school science clubs;
5. Organise career-mentoring public talks and lectures on science, engineering and technology through guest speakers who can dabble as role models;
6. Conduct science and technology educational and promotional video shows
7. Organize science weeks/festivals/camps
8. Orient school children and teachers to basic computer knowledge.

1.3 Justification

Poor public understanding of science and technology in Malawi is evidenced by low enrolment numbers in schools and poor examination results in science and technology related subjects. An analysis of the Malawi School Certificate of Education Examinations (MSCE) results in science subjects for a period 2002/03 to 2006/07 academic years reveals severe underperformance of students in the science subjects such as Physical Sciences, Biology, Agriculture and Mathematics (Table 13.1). Establishment of Science Discovery centres will therefore enhance the understanding of science amongst the youths and the elderly who will in turn encourage the youths to study/pursue science related subjects/courses.

Table 13.1: MSCE Science Subjects Pass Rates from 2003–2007

Year	*Sex*	*Subject Name*	*Total Number of Students who Sat for Exams*	*Total Number of Students who Passed Exams*	*Pass Rate (per cent)*
2003	F	agriculture	19367	10192	52.63
2003	F	biology	16125	9011	55.88
2003	F	mathematics	18924	7001	37
2003	F	physcience	1556	955	61.38
2003	M	agriculture	30598	22347	73.03
2003	M	biology	23611	16740	70.9
2003	M	mathematics	28121	14431	51.32
2003	M	physcience	2512	2070	82.4
2004	F	agriculture	22197	15770	71.05
2004	F	biology	20833	9743	46.77
2004	F	mathematics	21748	6907	31.76
2004	F	physcience	3738	2271	60.75
2004	M	agriculture	33972	28443	83.72
2004	M	biology	29322	18240	62.21
2004	M	mathematics	30716	14627	47.62
2004	M	physcience	4711	3858	81.89

Contd...

Table 13.1–Contd...

Year	*Sex*	*Subject Name*	*Total Number of Students who Sat for Exams*	*Total Number of Students who Passed Exams*	*Pass Rate (per cent)*
2005	F	agriculture	25472	16968	66.61
2005	F	biology	27602	18328	66.4
2005	F	mathematics	26431	10528	39.83
2005	F	physcience	9553	4303	45.04
2005	M	agriculture	38770	31881	82.23
2005	M	biology	38893	30421	78.22
2005	M	mathematics	36368	19235	52.89
2005	M	physcience	11165	7384	66.14
2006	F	agriculture	30352	14490	47.74
2006	F	biology	33176	11724	35.34
2006	F	mathematics	32443	9396	28.96
2006	F	physcience	14453	5522	38.21
2006	M	agriculture	43729	27796	63.56
2006	M	biology	44718	24912	55.71
2006	M	mathematics	42957	18161	42.28
2006	M	physcience	15462	9304	60.17
2007	F	agriculture	32692	9482	29
2007	F	biology	36545	7823	21.41
2007	F	mathematics	35655	15281	42.86
2007	F	physcience	18106	7883	43.54
2007	M	agriculture	48314	22372	46.31
2007	M	biology	51041	16758	32.83
2007	M	mathematics	48989	25716	52.49
2007	M	physcience	21072	12950	61.46

Malawi joins the rest of the world in recognizing the major role Science Centres play in the development of science, technology and innovation through raising public awareness of current research and stimulating an interest in science, among other things. Current state of affairs in the world indicates that science and technology play a pivotal role in socio-economic development processes of a country. For example, most of countries of South-East Asia, who at the time of Malawi's political independence (1964) were poor in terms of development, now have economies that have benefited extensively from determination of their governments to integrate science and technology in national development planning.

Although Malawi has a number of formal science and technology education and training institutions, these do little in terms of promoting public understanding of science, engineering and technology outside the classroom. This could be due to inadequate resources as a consequence these institutions concentrate on the primary objective of delivering formal science education and training. Establishment of Science Centres therefore, through their outreach activities and interactive science exhibits, would take the learning and understanding of science and technology out of the classroom to the members of the public.

2. Project Components

The project proposal had a number of components that require financing and these components have been described as follows:

2.1 isk@ppiptek.or.id

In order to successfully and sustainably run the activities of this science centre to the desired expectations and standards, the attainment of the requisite skills is very central. Thus, both short and long term training in relevant fields such as science communication, science and technology policy, and science education becomes paramount. This component will produce high caliber personnel that will be able to execute the activities of the science centre.

2.2 Construction of the Science Centre Building

The science centre building will include the following compartments;

Exhibition Floor

This is an exhibition area for displaying and operating interactive science exhibits;

Demonstration Laboratory

This is where science demonstration shows and experiments are performed by an educator. The laboratory shall have a safe store room where pieces of apparatus and stocks of various chemicals and other materials will be kept. It shall also have an experiment preparation area. The capacity of the demo lab shall ideally be fifty people;

Auditorium/Lecture Theatre

This shall be used for showing science and technology videos. It shall also be used for conducting science and technology talks and lectures;

Resource Centre

This shall stock a variety of science and technology books and magazines. It shall also house computers for visitors hands-on practicing and learning of computer based science activities/tasks. The resource centre shall also have Internet facility;

Maintenance Workshop

This is where scientific exhibits/models shall have to be manufactured and/or repaired. It shall house the necessary tools and equipment;

Kiosk Facilities

This will not only generate some income for the centre's activities but will also bring convenience to the visitors who shall be in need of snacks and drinks within the centre's vicinity thereby making them spend more time at the centre to interact with the exhibits.

Meeting/Conference Facilities

The centre will have a conference hall. This will be used for holding conferences/ workshops, public lectures and career talks.

2.3 Mobile Science Discovery Programme

Realizing the need for taking science and technology to the people and make them learn it from their own natural environment, this programme component will offer an opportunity to the public for outreach science and technology demonstrations, shows and exhibitions. The programme will ideally offset the cost that the audience might have been incurring through traveling to the science centre.

3. What is Currently being Done in the Absence of Science Centres?

As alternative sources of funding are being explored to establish Science Centres, informal science learning is currently being enhanced, amongst the youths, through the formation of school science clubs. These clubs visit Museums, Zoos, National Parks, Wildlife Reserves, Universities and other institutions that offer science courses and/or manufacturing industries.

Acknowledgements

I am grateful to members of staff of the National Research Council of Malawi for providing me with relevant information to come up with this write up.

Special thanks also go the Director for the Department of Science and Technology, Mr Henry Mbeza, for his support and technical advice.

References

Anderson, D. and Zhang, Z., 2003. Teacher perceptions of field-trip planning and implementation. Visitor Studies Today;

Bitgood, S., Serrell, B., Thompson, D., 1994. The impact of informal education on visitors to museums. In V. Crane, H. Nicholson, M. Chen and S. Bitgood (Eds.).

Freechoicescience education: How people learn science outside of school, New York: Teacher's College Press, pp.2-9

Chapter 14

Setting up Science Centres: Sri Lankan Experience

M.C.N. Jayasuriya

National Science Foundation, Sri Lanka

E-mail: dir@nsf.ac.lk, jayasuriya@nsf.ac.lk

ABSTRACT

Science centres are non-formal educational institutes nurturing interest in the life-long learning of science and technology. They fulfill an educational dimension that is often not provided for in the educational systems of most countries. Science Centres sometimes referred to as Science Parks are full of hands-on experiments that arouse the curiosity of both young and old. They are often fun for youngsters and have been found to be 'cool' for hard-to-please pre-teen school going children. Most developed countries have science centres, which are generally colourful, busy and buzzing with things to see and do. There are real life exhibits with large collections of specimens of interest, whether it be natural history, wild life, cultural or other. Sri Lanka has no science centre to boast about. It has a public museum under the Ministry of Cultural Affairs National Heritage and a so called 'Science Park' belonging to the National Engineering Research and Development Centre of the Ministry of Science and Technology. The National Engineering Research and Development Centre (NERD Centre) coming under the purview of the Ministry of Science and Technology has the only Science Park associated with the Ministry of Science and Technology. It is a Technology Park established with the objective of educating, inspiring and entertaining children and youth in particular, on history of local inventions and current developments in engineering science and technologies developed by the NERD Centre. It is also a display centre promoting awareness, understanding and appreciation of technology. The Ministry of Science and Technology is exploring the possibility of establishing a Science Centre with the main objective of popularizing science amongst school children and the general public. School children seem to be moving away from basic sciences because of the attractive prospects being offered by disciplines such as information technology, accounting and management. The proposed centre will have facilities for real life exhibits and hands-on-experiments that will attempt to satisfy the curiosity of both children and adults and be a medium of education and awareness creation in science and technology. This workshop should provide adequate information on basic principles in the establishment and day-to-day operation of a productive science centre for Sri Lanka.

Keywords: *Science centre, Science museum, Non-formal educational institutes, Real life exhibits, Hands-on-experiments.*

Introduction

Investment in science and technology is vital for economic and social development of any nation. Without such investment no country will develop to gain economic independence. One of the key factors responsible for the decrease in the development in Science and Technology in Sri Lanka has been the poor investment towards research and development. While in most developed and developing countries, there is an increase in the investment for R&D, in Sri Lanka the GERD/GDP (Government Expenditure for Research and Development/Gross Domestic Product) ratio has decreased steadily. The per cent of GDP spent on Research and Development has decreased gradually from 0.3 per cent in 1966 to 0.17 per cent in 2000. We have not allocated even 0.2 per cent of our total expenditure to R&D: agencies such as UNESCO have recommended that the minimum per cent of GDP that should be spent on R&D should be at least 1 per cent .

The public perception of science and technology in Sri Lanka is lack of interest and trust, perhaps associated with ignorance. Very little importance is given to scientific events both by electronic and print media. Majority of Sri Lankans including policy makers, government officials and the younger generation are yet to change their mindsets about the importance of science, technology and innovation. This is well reflected in the very poor investment in science and technology in the annual budgets and the keen interest shown by the young generation in opting for courses in non-science related subjects such as management and accountancy as against science, engineering or mathematics.

Science and Technology Parks/Science Centres/Science Parks are economic and technological development complexes aimed at fostering knowledge-based economies by bringing together scientific research, business and governmental organizations in one physical location, and supporting interrelationships between these groups. In addition to providing space for knowledge-based products, science and technology parks can house centres for scientific research, technological innovation and incubation, training, forecasting, as well as facilities for fairs, exhibitions and market development. They are formally linked (and usually physically close) to centres of technological excellence, universities and/or research centres.

The mission statements of science centres vary. But all depict them as places that make science accessible and encourage the excitement of discovery. They have become an integral and dynamic part of the learning process, promoting exploration, investigation and innovation.

Science Centres are also becoming a valuable part of recreational infrastructure. Silicon Valley (USA) is the pioneer in developing science parks in the world, Sophia Antipolis (France) in Europe and Tsukuba Science City (Japan) in Asia. Silicon Valley (originally Stanford University Science Park) was started in the early 1950s, Sophia Antipolis in 1960s and Tsukuba Science City, in early 1970s. They are the oldest and the most well-known science parks in the world. Today, there are over 400 science parks in the world and the number continues to grow. At the top of the list is the USA, which has been reported as having more than 150 science parks. It is followed by Japan with 111 science parks. China started its effort in developing science parks in the mid-1980s and now has around 100, of which 52 were approved by the national government and the rest are approved by local government.

Objectives

The primary objective of the proposed science centre in Sri Lanka is to provide the opportunity for school children and the general public to gain an insight into science and technology and an opportunity to experience various branches of science. The exhibits could take them to the age of the dinosaurs, wonders of nature, space exploration, climate change and natural disasters. They encourage people to think about what we can do to protect this planet and bring about the necessary balance and harmony to create the right climate for the future generations. At present only a small section of the public have a science back ground or have had a science education and access to scientific literature. Some may have information on scientific development through the media, which is often inadequately presented. The large majority of schools going children and adults have little or no access to scientific information that will arouse their interest.

In schools too the division of education into streams brings about a physical barrier for all students to learn about science and experience science. Only those in the 'science stream' will have the opportunity to study science and scientific methods and be involved in projects that may give an opportunity to obtain hands-on experience in scientific methodology. Others have very little exposure to science per se during their school education.

While children in the 'science streams' have the opportunity to study science they have very little if at all, facilities for gaining laboratory experience since most schools lack basic scientific laboratory facilities. Practical experience even at Advance level is minimal in most schools.

Science centres (sometimes called Science Museums, which is a misnomer) are non-formal educational institutes nurturing interest in the life-long learning of science and technology. They are centres fully devoted to science and fulfill an educational dimension that is often not provided for in the educational systems of most countries. The older Science Museums tended to concentrate on static displays of objects related to natural history, paleontology, geology, industry and industry related machinery. The modern Science Centres have broadened the range of subject matter and introduced many interactive exhibits. They are full of hands-on experiments, often giving attention to technology that will arouse the curiosity of both young and old.

Most developed countries have science centres, which are generally colourful, busy and buzzing with things to see and do. There are real life exhibits with large collections of specimens of interest, whether it be natural history, wild life, cultural or other.

Scope

Science centres have to be different from museums both in concept and in operation. While a museum has static exhibits of historical value and importance a science centre will be a highly interactive operation that would demonstrate principles, present the existing state of the art and predict future potential of the various branches of sciences. It would be more oriented to the future than the past. The educational emphasis would be to promote science and attract students to sciences and encourage the development of a scientific culture. There will be more working models providing the visitor with the opportunity of learning through participation and interaction.

Many areas of science could be included in these centres. The final product will be dependent on the requirement of the country and the emphasis being given to specific subject areas.

While some areas such as principles governing various sciences, human physiology, information technology, space exploration, environmental sciences etc. are common to most countries, indigenous knowledge could be very specific to individual environments. The composition will therefore depend on priorities and the availability of financial resources.

Priority given in the establishment of the centres would be for participation and experimentation. These could be visual based (video programmes) but more importantly interactive. Where such interactive experimentation is not possible models, illustrations and demonstrations could provide the basis of explanation. While it would be important to have guidance by way of 'official guides' the students themselves must be allowed to 'experience' and obtain 'first hand knowledge' on the relevant subject.

The location, buildings and the entire infrastructure requirement will have to be determined by the financial resources available for establishment and maintenance of the centres. However, in order to be able to serve the entire country it should be centrally located with easy access to general services such as adequate water and undisrupted electricity supply as well as back up services and perhaps in close proximity to the scientific community.

Present Status

Sri Lanka has no science centres at present. It has a public museum under the Ministry of Cultural Affairs National Heritage and a so called 'Science Park' belonging to the National Engineering Research and Development Centre of the Ministry of Science and Technology.

The National Museum established in 1877 began as a place for displaying cultural and natural heritage of the country. Several wings were added to the main building from time to time facilitating the extension of the library, ethnological and anthropological studies. The installation of the Throne and crown of the last king of

Kandy which were handed over back to Sri Lanka by the British Government, has greatly enhanced the quality of the museum collection. The thematic arrangement of the galleries has provided an opportunity for visitors to study the ancient culture of Sri Lanka under one roof. The museum has a Science section with a large number of specimens in entomology (Figure 14.1), zoology and botany (Figures 14.2 and 14.3), Geology (Figure 14.4), a Cultural section with over a million exhibits of coins, jewellery, pottery etc., a Conservation section and a Library. The museum now has nine branches, spread out in the provinces, a school science programme and a fully fledged mobile service.

Figure 14.1: A Collection of Over 95000 Dry Mounted Insect Specimens in Storage Cabinets and Over 5000 Wet Preserved Insect Specimens in Alcohol (Courtsey www. museum.gov.lk)

Figure 14.2: Collection of Fauna Indicating its Diversity in the Country (Courtsey www. museum.gov.lk)

Figure 14.3: A Collection of Botanical Specimens Comprising of Over 1500 Dried Plant Material and a Small Wet Preserved Collection (Courtsey www. museum.gov.lk)

Figure 14.4: The Geology Division Maintains a Collection of Minerals, Soils, Fossils and Prehistoric Items (Courtsey www. museum.gov.lk)

Science Section

The National Engineering Research & Development Centre (NERD Centre) coming under the purview of the Ministry of Science and Technology has the only Science Park (Figure 14.5) associated with the Ministry of Science and Technology. It is a Technology Park established in 1999 to celebrate the Silver jubilee of the NERD Centre. The objective of the park is to educate, inspire and entertain children and youth in particular, on the history of local inventions and current developments in engineering science and technologies developed by the NERD Centre. It is also a display centre promoting awareness, understanding and appreciation of technology.

Figure 14.5: Technology Park of the NERD Centre Showing Current Developments in Engineering and Technologies of the Centre (Courtsey www. museum.gov.lk)

Conclusion

A Science centre for Sri Lanka has been under consideration for the past 50 years. But various obstacles, especially financial constraints have been responsible for the delay in its establishment. The need for a well planned science centre for Sri Lanka is continuingly and strongly felt. Many developing countries in Asia have science centres of high sophistication. Korea and Malaysia rank high in this regard. It is hoped that a high caliber science centre would provide the Sri Lankan children and the general public the necessary awareness in science and technology and give them the opportunity to become scientifically literate and forward thinking citizens and beneficiaries of universal scientific knowledge.

Acknowledgements

The pictures presented here were taken from the websites of the National Museum of Colombo and the National Engineering Research and Development Centre of the Ministry of Science and Technology.

Chapter 15

Investment in Science, Technology and Innovation

Gwakisa Bapala

Ministry of Communication, Science and Technology, Tanzania

E-mail: gwakisa.bapala@gmail.com, bapala@msthe.go.tz, bapala2006@yahoo.co.uk

ABSTRACT

The focus of this paper is to take stock of the status and opportunities for investing in Science and Technology in Tanzania. The paper prefaces this discussion by reflecting on an apparent paradox. While Science, Technology and Innovation has never been better studied and understood, many of our ideas about science, technology and innovation have failed to fundamentally change the institutional and policy setting of public and private investment intended to promote these tools for development. The paper surveys the status and what is available with regards to science, technology and innovation in Tanzania and the contribution it can provide for social and economic development. The paper then locates the current interest about science, technology and innovation systems in the evolving and contested approaches to development, noting that this is characterized by a long history of false dichotomies. The contingencies of new and emerging technologies will demand more networked modes of collective intelligence and innovation that are embodied in the science, technology and innovation. The paper argues, however, that the science, technology innovation status in Tanzania should be viewed as an opportunity for more investment, rather than another competing narrative. For the way forward, it is suggested, to create international collaboration on different areas of science, technology and innovation such as science centres etc. to kick-start the virtuous spiral of innovation practice and policy learning. This is needed to strengthen Science, Technology and Innovation developmental goals. The paper argues that it is the responsibility of all of us to argue for this space for diversity to flourish and to help consolidate and promote what is known about Science, Technology and Innovation. If we are not more successful in stimulating institutional and policy change we will still be debating these issues even in the future.

Keywords: *Science, Technology, Innovation, Science centres, Research and development*

Introduction

Status and opportunities for investing in Science and Technology in Tanzania with some reflection on the question of "what is available when you talk of science, technology and innovation in Tanzania" and use this reflection as a lens to look at the challenges toward strengthening of science, technology and innovation systems.

Historically science, technology and innovation in Tanzania have been known, perceived and taken in policy issues as a priority since 1960. Tanzania is one of the few countries in Africa that implemented UNESCO's recommendation of the 1960's to establish a national science and technology policy. (United Republic of Tanzania [URT] 2003 S&T Sub Master plan). Various series of events and steps have been followed up by the Government of Tanzania to implement the recommendations.

Tanzania resulted from the merger between Tanganyika and Zanzibar in 1964 into a United Republic of Tanzania. It is located between the latitudes 1°S and 12°S and the longitudes 30°E and 40°E and covering nearly one million (945,087) sq.km of mostly arable land. The topography and climate are very variable across this mainly plateau country. It harbours abundant natural resources namely hydropower potential, tin, phosphates, iron ore, coal, diamonds, gold, natural gas, nickel and various attractive tourist features such as natural animal reserves, forests, mountains which are capable of supporting major development programmes so as to improve the quality of life of its presently nearly 36 million inhabitants with literacy rate of 68.7 per cent.

The average per capita income in Tanzania is US $ 0.70 per day; the economy is growing at a rate of 3.5 per cent. Agriculture is the largest productive sector (46 per cent) while the manufacturing industry contributes to less than 10 per cent of our GDP; while agriculture enjoys on the most important sector in terms of production employment and export earnings, its contribution has depended on a small number of crops whose value has been adversely affected by various international commodity prices.

Materials and Methodologies

Evolution of Science, Technology and Innovation in Tanzania

In spite of the fact that there are a lot of constraints on the availability of resources, a lot have been done by the Government in terms of investing in science, technology and innovation in the country. This has been proved by the formulation of policies, acts and their related commission to fore-see and implements these policies and acts. Specifically the following were done immediately after the perception that science, technology and innovation is the key for socio-economic development.

1968/72 A desk was established at UNESCO National commission, which by then was under the Ministry of Education. This desk was responsible for Science and Technology in the country. The parliament enacted Act No 5 that establishes Tanzania National Scientific Research Council (TNSRC) it was under the Ministry, responsible for National Education. The official establishment of TNSRC moved from the Ministry of Education to the Ministry of Planning and Economic Affairs.

Formulation of National Science and Technology Policy: Transformation of TNSRC into Act No 7, which establishes Commission of Science and Technology.

Establishment of the Ministry's that is responsible for Science Technology and Higher Education.

Review of Science and Technology Policy in Tanzania, 2007

Institutional Framework for Science and Technology

The department of Science and Technology is within the Ministry of Higher Education, Science and Technology. This is a Cabinet Ministry that is responsible for science and technology, including higher education. All matters relating to science and technology are operated through the Directorate of Science and Technology under the Permanent Secretary and hence to the highest organs in the government democracy. Together with other line departments which are Higher Education and Technical Education, it is responsible for formulation, review, monitoring and evaluation of science and technology policies in the country.

Department of Science and Technology

Specifically the Department of Science and Technology is responsible for;

- Developing comprehensive, up to date and sustainable national science and technology policy with an adequate co-ordination, framework in terms of the legal, institutional, human, physical technical and financial resource dimensions;
- Creating a conducive environment for implementation, monitoring, evaluation and transfer to target end-users priority strategic national science and technology programmers designed to acquire, develop and or transfer new scientific and technical knowledge and innovations;
- Working closely with other national stakeholders to mainstream science and technology in the national educational system at all levels and in the cultural fabric of the Tanzanian society so as to ensure its effective application in poverty eradication.

The Tanzania Commission for Science and Technology (COSTECH)

The Tanzania Commission for Science and Technology (COSTECH) is an apex semi autonomous executive government agency under the Ministry of Science Technology and Higher Education. The Commission brings together the top leadership of the scientific and technological institutions in the country under one forum. Thus, the Commission maintains a system of collaboration, consultation and cooperation with parties within and outside Tanzania whose functions relate to the application of science and technology for development.

The Commission is entrusted with the responsibility of coordinating and promoting science and technology development activities in the country. It is the principal advisor to the Government on all matters pertaining to Science and technology and their application for socio-economic development of the country.

Science, Technology and Innovation in Tanzania

From historical point of view, it proves that Tanzania has tried its best as far as investment in Science, Technology and Innovation are concerned. However, the results towards this are tremendously low. Like other socio-economic infrastructures, science and technology in Tanzania is still not well developed, primarily because of the low level of scientific research activities (Nkunya, 2003). Efforts have been made by both Public and Private sectors in ensuring that various Science, Technology and Innovations (STI) thematic areas are harnessed and utilized fully so that they contribute to socio-economic development. These include:

Research and Development

Research plays a very crucial role in the socio-economic development of any society. The main goal of research is to serve as an instrument of change to improve people's living standards by stimulating growth and increased productivity in critical productive sectors of the economy. It can lead to spin-off of new firms, creation of more competitive products and services in a country for the improvement of quality of people's life thereby increasing life expectancy, enhanced food productivity and security, shelter and world markets, among many other benefits. At the level of enterprises, research can bring about product innovations, product improvement, increased service efficiency, effectiveness and improved performance in the market place.

There are Public and Private Institutions which are conducting Research and Development for various reasons. Some of the reasons include, undertaking to be just for acquisition of work related promotions to researchers and academicians and others are being driven by donor's interest rather than societal problems. Currently institutions are scattered over many ministries, and few are attached to higher education institutions especially Universities (S&T sub master plan 2003).

In spite of the fact that the sector is not well coordinated, all major national research and development institutes and institutions of higher learning are affiliated to the Tanzania Commission for Science and Technology (COSTECH). The advisory committee under COSTECH operates as a thinktank to advise the Commission on sectoral priorities for research, allocation of resources, review of research proposals and initiate recommendations for review of the National Science and Technology Policy. The institutions arranged against major R&D areas include, Industry and Energy Research, Food and Agricultural Research, Natural Resources Research and Energy Research.

Most Universities and institutions in Tanzania are prioritizing research undertaking due to its importance for socio-economic growth. They have also included research in basic science among the priorities in their daily undertaking as it comes to. These are being provided by most of the research policy and agenda which as the matter of fact they are at the Institution level. Due to scarcity of resources sometimes it becomes difficult to implement the Institution's stipulated research agenda. To solve this problem, the Government of the United Republic of Tanzania (URT) has decided to formulate the National Research policy which at this particular moment has already been discussed by the stakeholders, waiting for the government machinery to be put it in force. The National Research policy amongst others will address critically the issues related to:

National Research Agenda

National development needs ought to be harmonized and prioritized at societal level so that national research teams could be formed and funded centrally in pursuit of nationally perceived solutions. Most research activities are apparently driven by the agenda of various funding sources, be it external or local. Such activities have to adequately address the country's development agenda and the needs of society. This therefore calls for establishment of multi-disciplinary research teams that would prioritize National Research Agenda for short-, medium- and long-term articulation and which will be reviewed in every five years.

Research Funding

Funding is one of the critical components for successful and effective research activities in a country. However, a research undertaking is expensive requiring heavy investment for a longer time before its impacts are appreciated by society. In order to be able to address issues of national priorities, allocation of funds for research from internal sources is of paramount importance. Funding of research is one of the major problems in the country due to the very limited financial resources that are available to research and development institutions and to individual researchers. In 2003–04, research funding amounted to about 0.012 per cent of GDP as compared to 1 per cent of GDP proposed in the 1996 National Science and Technology policy and SADC countries. The above figure is far below that of South Africa, which was 0.81 per cent and 2.76 per cent in the United States of America for the same year. Thus, a holistic funding mechanism for research is necessary for sustained research effort. A holistic approach entails a well-coordinated research fund facility that takes on board national priority areas with equitable access.

Human Resource Base

There is inadequate human resource capacity in most fields in terms of both quality and quantity for effective participation in the national research activities. Available data indicates that there were only 13 research personnel for every 10,000 persons in Tanzania in year 2006 compared to 1 researcher for every 1000 members of the workforce in South Africa, 5 in Australia and 10 in Japan. There is also limited capability and experience in converting research results into useful processes, products and services.

National Research Support Environment

Research activities are slow in producing tangible results. When subjected to resource-constrained environment, there is tendency to marginalize its importance and support from the Government and public authorities as well as the private sector. With continued marginalization, development opportunities emanating from research results can never be realized. There is need therefore to create a conducive national research environment to ensure that national development is research driven.

A Comprehensive Research Management System

The existing research management mechanisms in place are inadequate for effective and sustained planning, monitoring and evaluating research activities in the country. Further, research activities have to be inclusive and the research results adequately documented and publicized. A comprehensive research management system is necessary to ensure well-focused and effective research that addresses national priority needs.

Partnership Between Research Institutions and Industry

Effective application of research results requires strong linkage between research institutions and industry. There is a need to strengthen partnership between research institutions and the private sector (industry) in development of joint research programmes with practical application to the technical and business problems facing the nation/industry/private sector. Socio-economic development depends to a large extent on aligning the flow of knowledge to knowledge-supply chains aimed at enhancing value of processes, products and services. There is need therefore to focus on demand-driven research and to encourage researcher interaction with economic situations and problem-solving alternatives.

Multi-disciplinary in Research

Most societal challenges are multi-disciplinary in nature and hence researches aimed at addressing these challenges have therefore to be multi-disciplinary in approach. It is important that there is a cross-disciplinary collaboration or multiple cultural perspectives that require more than one skill. Experiences from other countries have shown that most renowned research results emerge from cross-disciplinary collaboration or multiple cultural perspectives that requires more than one skill to realistically solve problems. There is therefore a need for different institutions and individual researchers to collaborate in research including commercialization of the same.

Environment for Protection of Intellectual Property Rights (IPR) and Patents

Limited awareness of IPR and the absence of an enabling environment including the non-use of patents in research activities among academicians and industrialists affect research in terms of its economic contributions and the value added in new products and processes, licenses with royalties. Through effective use of IPR, ideas that emerge from research carried at universities and other research institutions, which are ripe for commercial exploitation, ought to be allowed to develop and flourish in a protected environment before being subjected to the rigours of full competition. If the outcome is successful, it is imperative that any IPR should be adequately protected before commercial exploitation.

Regional and International Cooperation and Partnerships

There is limited exploitation of opportunities emanating from various regional and international collaborative frameworks. These collaborations range from regional to bilateral and include international organizations and agencies. For example, the Government has benefited much on capacity building from bilateral cooperation with the Swiss Government and other multilateral organizations such as the SIDA-SAREC, World Bank, African Institute for Capacity Development (AICAD), Rockefeller Foundation, International Atomic Energy Agency (IAEA), the Comprehensive Nuclear Test-Ban Treaty Organization (CTBTO) and the International Centre for Genetic Engineering and Biotechnology (ICGEB). However, much more co-operation is yet to be evidenced with many other regional and international organizations in order for the nation to reap maximum benefits from such collaborations.

Mainstreaming of Environmental Concerns

Research activities need to be compliant with the environmental laws and regulations in the country. There is a need to mainstream environmental concerns in all research activities in the country.

Information Management System

Undergoing Research requires strong information management system accompanied by an effective and efficient ICT infrastructure. There is a need to have adequate database on research activities, inputs, outputs and impacts to the society; mechanism for dissemination of research outputs to the society, sharing information by research institutions, individual researchers and other bodies on research programmes. Investment in ICT infrastructure at micro- and macro-levels, coupled with adequate levels of human resources–in terms of numbers and qualifications–to utilize ICT in research are basic conditions for a conducive research environment. Researchers should have access to most journals available online to facilitate research at reasonable cost of connectivity. High-speed bandwidth/broadband as well as researcher skills in searching for the right information and sources are of prime importance.

Research Performance and Quality Assurance System

To ensure research effectiveness, performance monitoring and quality assurance system has to be put in place. There exists neither institutional nor national mechanism for assessing research performance by stakeholders in the country. There is a need to develop research performance indicators aimed at ensuring quantifiable outputs in terms of how research addresses issues relevant to national or societal needs as well as that of advancement of academic knowledge or the state of knowledge.

Research Ethics

Conduct of research requires a sound ethical framework. Research activities need to have therefore strong scientific merits in terms of justification/rationale for conducting it; soundness, clarity of the objectives and methodologies to be employed.

There should also be equitable sharing of research benefits, particularly between foreign and local researchers / institutions / research communities.

Proper Institutional Memoranda of Understanding (MoU) between collaborating institutions / countries, effective review committees at the institutional level; legal frameworks such as IP policies and Materials Transfer Agreements (MTAs) at both institutional and national levels are necessary in ensuring and enforcing adherence to research ethics.

It is now a high time for the implementation of the ideas of establishing a strong linkage between Universities / centres with business / entrepreneurship, that is, when the concept and need of establishing an effective and efficient Science Centres comes in. Science Centres can play a very important role that may lead into spinning-off of new firms, creation of more competitive products and services in a country for the improvement of quality of people's life thereby increasing life expectancy, enhanced food productivity and security, shelter and world markets, among many other benefits. At the level of enterprises, research can bring about product innovations, product improvement, increased service efficiency, effectiveness and improved performance in the market place.

Various sectors have R&D technologies or products that need to be commercialized. Example of such sectors and associated technologies are explained in the Table 15.1.

Table 15.1: Selected Research Based Products/Processes and Technologies Developed by Different Sectors in Tanzania

Sl.No.	*Name of Relevant Sectors*	*Types of Technologies/Products*
1.	Agricultural and Food Security	• Improved Charcoal stoves • Oil expellers • Mushroom cultivation technologies • Bio-diesel technology • New seed varieties for maize, beans, oranges, mangoes, tomatoes, • Food processing machinery
2.	Livestock	• New species of cattle, • Improved species of milking goats and cattle
3.	Industrial	• Glue from cashew nut shell • Water pumps • Fruit juice processing equipment • Cassava processing equipment • Motorized hand driven tractor • Mining and mineral processing machinery • Construction equipment

Contd...

Table 15.1–Contd...

Sl.No.	Name of Relevant Sectors	Types of Technologies/Products
4.	Health	• Medical Incinerator • New drugs from indigenous plants for treating skin diseases • New Treatment for malaria • Mixtures for treating cough, asthma and peptic ulcers
5.	Education	• School chalk production • Production of sodium hydroxide for science laboratories
6.	Energy	• Solar energy, wind energy • Energy from agricultural wastes • Dryers and furnaces

Source: Draft STI Policy.

Science Education

Education is a long-term intellectual change and adaptation process. Educational attainment encourages aspiration and self belief in the next generation–to compete in the global economy, to live in a civilized society and to develop the talents of each and every one of us (UNESCO 2002 pg 96). For the better application of Science and effective utilization of Science Centres we must increase the number of students who will be enrolled in Science subject by including even the private sectors. Since 1995, the public/private blend has continued to dominate secondary provision at a ratio of roughly 70:30. (Malyamkono and Manson)

With respect to student enrolment in science subjects in schools, progress in Tanzania, as in some of the other eastern-southern African countries, has not been as dramatic. Slow progress in this area has been principally at all levels of educational provision particularly at secondary and tertiary levels (Ishumi, 1994). Exception can be made of the primary school level, where, in a range from eight (before 1998) to six (after 1998) core subjects, the syllabus has always included Science and Arithmetic as basics not only for learning science and technology but presumably also for acquiring a scientific culture and spreading into the wider community. Problems of enrolment for learning of science start, historically, with secondary schools, where mathematics, along with the science subjects, especially chemistry and physics, have often been shunned by students, often with fears and excuses that these subjects are difficult (Meena *et al.*, 1987). This has been worsened by expansion of private secondary schools, most of which are arts and humanities based mainly due to the often less investments necessary. The result has often been a lopsided intake and enrolment in favour of the arts/humanities/social sciences. At the tertiary level, particularly at university entry point, the annual undergraduate student intake into the natural sciences for the whole of the first two decades of independence, was, on average, only a half the figures for the liberal arts (the social sciences, arts and humanities with corresponding graduate outputs running similarly at half the rate (Ishumi, 1994).

Table 15.2: Undergraduate Student Enrolment into the Sciences vis-à-vis the Humanities at the University of Dar es Salaam, 1966/67–1984/85

Year	*66/67*	*67/68*	*68/69*	*70/71*	*72/73*	*74/75*	*76/77*	*78/79*	*80/81*	*82/83*	*84/85*
Sciences	-*	-*	-*	368	471	372	462	378	395	406	427
Arts/Humanities	565	738	922	1,032	1,015	824	885	885	793	862	952
Prop of Science	0	0	0	0.4	0.5	0.5	0.5	0.5	0.5	0.5	0.4
Average ratio						0.5					

Source: University of Dar es Salaam Admission Office; (Ishumi, 1994).

* The first group of students to enrol at the University of Dar es Salaam were for the Faculty of Law; the second group were for the Faculty of Arts and Social Sciences in 1963; and the third group were for the Faculty of Science in 1964. UDSM remained the only university in Tanzania until 1984, when a second university, Sokoine University of Agriculture, was established with a hiving of UDSM's Faculty of Agriculture off to a new university.

At the tertiary level, particularly at university entry point, the annual undergraduate student intake into the natural sciences for the whole of the first two decades of independence, was, on average, only a half the figures for the liberal arts (the social sciences, arts and humanities).

This relative imbalance has since continued to characterize university education in the country, worsening in the more recent years–with an average of only 10.4 per cent getting into the sciences.

Science and Technology Centre and Centre of Excellence

Science

Science can be seen both as a content and as well as a method. As a content, science consists of knowledge and experiences that carries credibility on the basis of its correctness, consistency, testability, durability and reliability for use by humankind. As a method, science consists of "a way of knowing," *i.e.* a procedure or set of procedures established or adopted for arriving at results (Popper, 1961; Lakatos, 1978).

Technology

Technology is commonly perceived to be a constitution of physical tools used to transform resources, such as technical hardware, into new products or results. It is, in other words, the application of concepts that have evolved from science into situations of practical application for production of goods, artifacts or services aimed at bettering and advancing conditions of life and sustenance of humankind in time and place, for instance in a household, in a community or elsewhere within the wider society (Cardwell, 1957; Lall, 1997).

Innovation

This is a successful creation, development and marketing of new goods or successful application of new techniques or ways of working that improve the effectiveness of an individual and organization (Archibugi *et al.*, 1994). In other words, innovation must bring about measurable improvements in productivity, product quality and sales (Oyeyinka, 2004).

Science and Technological Centres are very few in Tanzania and those few are less effective. Many reasons have been said toward this kind of situation, these ranges from lack of public awareness and appreciation on the role of Science for socio-economic development, lack of capacity to create and formulate effective policies which will accelerate the establishment of the Science Centres, very few resources are allocated to the Science Centres to donor's interest are not in favour of the sustainable development of these Science Centres.

The Role of Indicators in the Development of Science, Technology and Innovation in Tanzania

Indicators are not only important for activities but for many others that cannot be measured directly. They are selected on the basis of plausible assumptions about their relationships with the variables they represent. It is therefore a tool to be used to assist in the assessment of some activities or consequences. Typically, assessment or evaluation makes use of the sets of indicators each of which may display or highlight some facets of the object under evaluation. S&T indicators are therefore indispensable S&T policy tools. Specifically, they fulfil the following major functions:

- Signalling or monitoring: giving insights into and drawing attention to the development and trends in the system and its environment, a function without which it is very difficult to make any firm decision and planning on;
- Accountability, evaluation and allocations setting and justifying budgets and giving insights into the performance of the system against the goals established by policy makers and planners;
- Legitimisation: support for existing policies;
- Awareness: providing information to set aside prejudices and incorrect perception of the performance of the system (CCST 1999).
- Trends in STI Expenditure in Tanzania
- Tanzania's Gross expenditure on per GDP Ratio (GERD/GDP)

The R&D Intensity, (GERD/GDP ration) has been computed using two GDPs, the current PPP GDP and the constant 1995 GDP. It is noted that Tanzania experienced a sharp increase in R&D expenditure intensity (GERD/GDP) between 1996 and 1998 with the ratio (PPP) rising from 0.0006 to a peak 0.0215, since then the ratio has been declining over the time period to reach 0.0018.

The absolute expenditure to by performance sector and by source of funding is presented below. These expenditures include money spent on research activities, capital expenditure and recurrent costs.

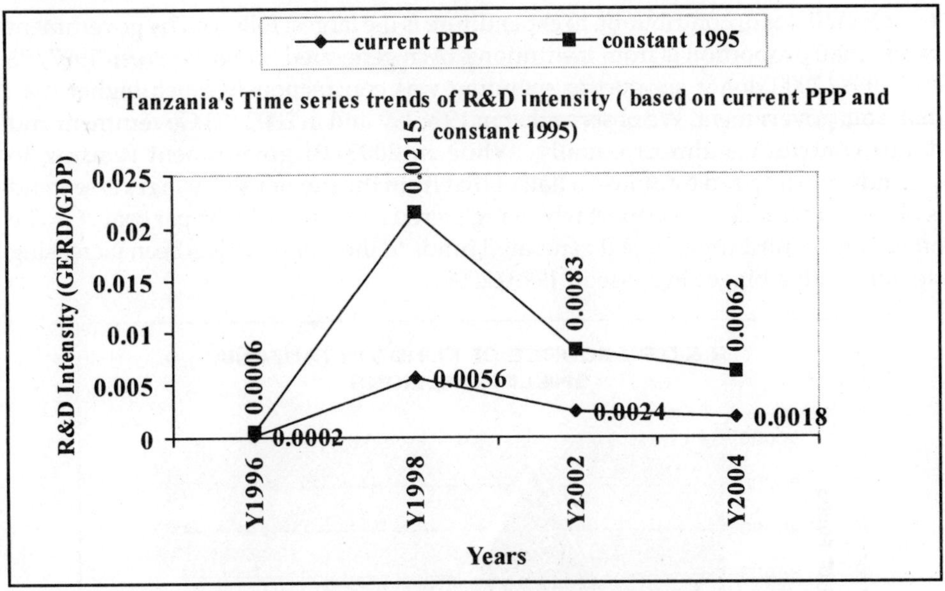

Figure 15.1: Tanzania's Gross Expenditure on R&D per GDP Ratio (GERD/GDP)
***Source*: A Survey of Tanzania's R&D inputs in government institutions 1995/96–2003/04**

Table 15.3: R&D Expenditure by Sector in Tanzania Shilling Millions

	95/96	*96/97*	*97/98*	*98/99*	*99/00*	*2000/01*	*01/02*	*02/03*	*03/04*
GOV	1403.7	11918.7	3867	7359.4	5521	7825.6	9198.5	11465.9	8359.4
HES	58.5	497.5	10111	8929.5	33060	33547	37099	39162.5	4418
SCS	26.3	66.6	78.2	198.9	211.7	495.4	480.2	681.4	804.3

Sources: A Survey of Tanzania's R&D inputs in government institutions 1995/96–2003/04.

We observe that most investment in expenditure goes to the Higher education (HE) sector followed by Government institutions while COSTECH receives a very tiny fraction. While higher education growth has been on a steady increase until 2004 when the growth rate declined, that of Government has been fluctuating mildly. The HE spending has increased due to establishment of new HE institutions and expansion. A strategy is needed to stimulate growth in R&D institutions and in COSTECH.

Table 15.4: R&D by Source of Funds in Tanzania Shilling Millions

	95/96	*96/97*	*97/98*	*98/99*	*99/00*	*2000/01*	*01/02*	*02/03*	*03/04*
GOV	3421.5	2870	4806.5	6814	9050	13400	12600	9070	4240
DONOR	891.8	411.5	6941.2	46042	5290	1760	11600	9070	19700
SELF	1271.9	1309	1329	3907	9580	1340	2219	2820	2930

Source: A Survey of Tanzania's R&D inputs in government institutions 1995/96–2003/04.

Overall donor contribution to expenditure is the largest followed by government and a small proportion is from institutions' own generated funds. Between 1997/98 and 1999/2000 donor support to spending was conspicuously much higher than that from government. We observe that in 1996/97 and in 2002/03 Government and donor contributes almost equally. While in 2003/04 government funding to expenditure dropped by almost a half of that from the previous year, a reverse trend is observed with donor support which registered an increase by 54 per cent . On the other hand contribution by self generated funds to though small has been increasing steadily with a bigger increase in 1999/2000.

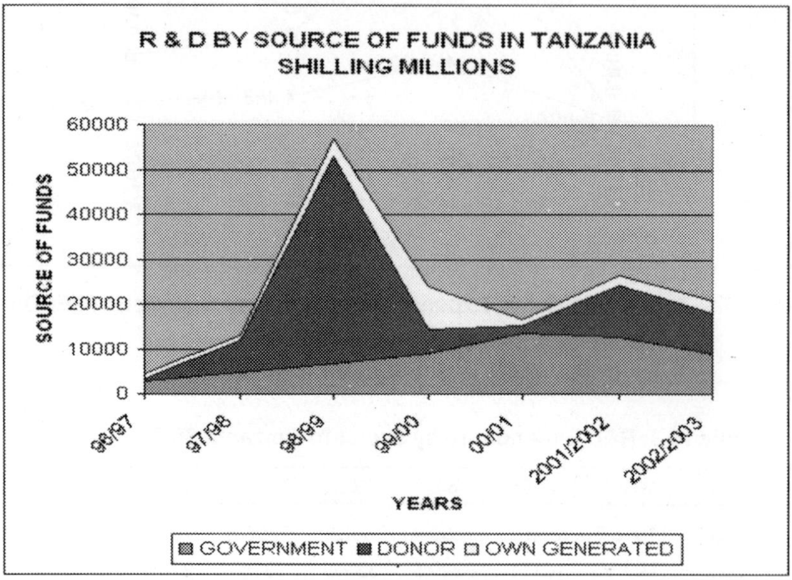

Figure 15.2

Trends in Tanzania Human Resources Inputs

Indicators commonly used for this R&D effort are the personnel Head Count (HC) and the Full-Time-Equivalent. Data on financial inputs over the 9 year time period though difficult to extract, have been relatively easier to get than data on personnel over the same period. More time is needed by institutions to extract information from files.

While data on HC for 2003/04 have been provided with ease, FTE data for the same year need computations and follow-up questions to respondents than time for this report could allow. As such the report has dwelt only on R&D personnel by Head Count. Also to be noted is that some institutions have not been able to provide such information, as such the reported results are not 100 per cent, they are, may be, about 85 per cent–90 per cent . Even then, we feel these numbers can still shed some light on the distribution of personnel in R&D institutions by performance sector, sex and occupation. Also important but not provided, is the age distribution by occupation and sector. We make comparisons with USA and South Africa.

R&D Personnel HC by Sector, Sex and Occupation as at 30th June 2004

Table 15.5: R&D Personnel HC by Sector, Sex and Occupation as at 30th June 2004

	Researcher		*Technical Staff*		*Others*		*Total*	
	Male	*Female*	*Male*	*Female*	*Male*	*Female*	*Male*	*Female*
GOV	464	207	431	265	428	296	1323	768
HES	383	102	23	2	160	59	566	163
SCS	67	21	34	11	37	34	138	66
	914	330	488	278	625	389	2027	997

Source: A Survey of Tanzania's R&D inputs in government institutions 1995/96–2003/04.

The numbers presented here do not include all personnel in the surveyed performance sectors. Some institutions have not been able to supply such information. Even then we still get the idea of the distribution of human resources in the R&D institutions by occupation and sex as can be seen in Table 15.4 and Figures 15.3–15.4.

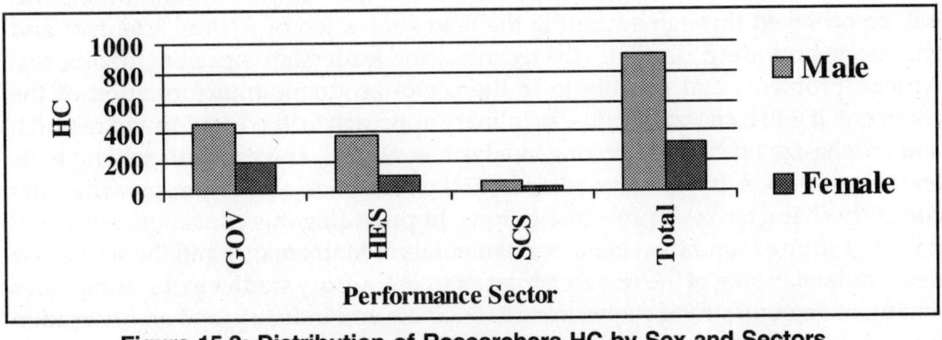

Figure 15.3: Distribution of Researchers HC by Sex and Sectors
***Source*: A Survey of Tanzania's R&D inputs in government institutions 1995/96–2003/04**

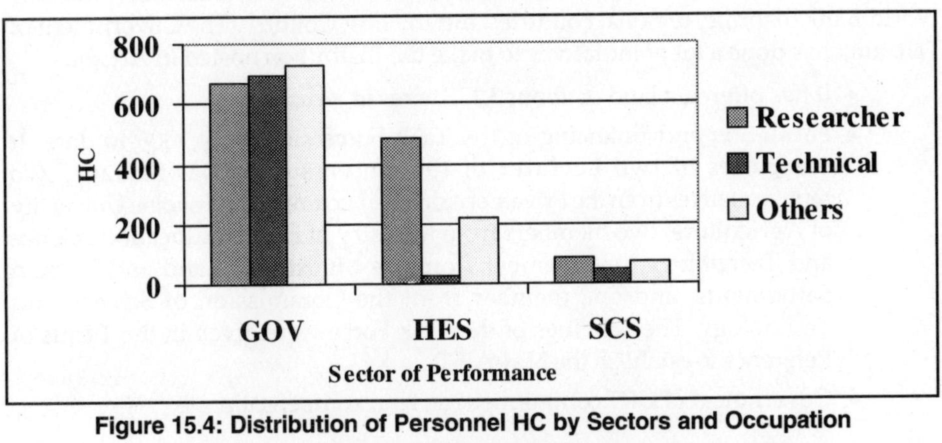

Figure 15.4: Distribution of Personnel HC by Sectors and Occupation
***Source*: A Survey of Tanzania's R&D inputs in government institutions 1995/96–2003/04**

Opportunity for Investment in Science, Technology and Innovations (STI) in Tanzania

The Government of the United Republic of Tanzania shall undertake to rise local funding for S&T research and development in order to attain the recommended minimum level of 1 per cent GDP at the earliest but in line with the adopted S&T sub master Plan. (S&T sub master plan 2003 pg 58). These efforts include:

Nelson Mandela Centre of Excellence

Tanzania was chosen by the advisory body of the Nelson Mandela African Institute of Science and Technology (NM-AIST) to host one of the four campuses in sub-Saharan Africa that will offer academic programmes with a focus on education and research in science and engineering to train future leaders, researchers and entrepreneurs. It will also attract the best and brightest faculty and students, selected on merit alone within PANAFRICAN member Countries. Currently there are two campuses already at the final stage of being established, these are the NM-AIST Abuja and Arusha Campus for the Western and eastern respectively.

The fundamental objective of NM-AIST is to develop Africa's human capital by producing world-class scientists, creating access, and mitigating brain drain. This will be achieved through educating the next generation of African scientists and engineers–providing them with the technical and leadership capacities to solve real African problems and contribute to the socio- economic transformation of the continent. It will focus on a multi-disciplinary approach to its education and research and emphasize on critical thinking, analysis, synthesis, and problem solving in its teaching methods. It will move away from the theoretical and rote approaches that currently characterize Africa's institutions. In providing this education, AIST will deliver a strong foundation in the fundamentals of Mathematics and the sciences; a deep understanding of the research frontiers; compulsory studies in the humanities and management; strong connections to business and industry; and an orientation towards entrepreneurship and service. AIST will emphasize links among academic disciplines and keep the barriers between these disciplines as low as possible.

The Institute will be operated in collaboration with NMI board of Directors, World Bank Institute, regional countries and the host Country. The Government of Tanzania has done a lot of initiatives to make the Institution hosted in Arusha.

- It has offered a land of about 3,125 acres in Arusha.
- Formation and financing of the Task Force since July 2006 to date. It comprises of two Lecturer of the University of Dar Salaam, two representatives from the Private sector, one Lecturer from Sokoine University of Agriculture, two members from Ministry of Higher Education Science and Technology, one member from the Ministry of Land and Human Settlements, and one member from the Commission of Science and Technology. The activities of the Task Force were given in the Terms of Reference to establish the NM-AIST.
- Government of URT commitment on land compensation.

- Government of URT has offered a temporary host of the Institute that is CAMARTEC.

East African member states are among the targeted Countries, and are therefore going to benefit directly from the existence of this Institution.

Sullivan Summit

One in three professionals trained in an AU country now lives outside of the continent (Owen Barder *et al.*, Report of the Centre for Development Working Group). The URT has realized this and therefore included it as a strategy to utilize ideas of the Diaspora in STI related matter. Sullivan Summit for 2008 would be held in Tanzania from 2nd June–6th June 2008 in Arusha. This is expected to be a room for about more than fifty scientists of which forty are from African origin and the rest are from other origins who are living in USA will interact and found out what will be the best platform for Tanzania to excel in Science, Technology and Innovation. The June summit will be a turning point as well as a break through on the way we perceive Research, intellectual Property etc and possibilities for attracting and linking them with member states. URT believe that a lot can be contributed by Diaspora involvements in professional development hence social economic development. Recommendation 20. The International community, other developing nations, and African Diaspora have the potential to play a critical collaborative role in Africa's economic development and technological capacity building. African Regional innovation communities should facilitate North-South and South- South collaborations as well as mobilize the knowledge net work of Diaspora for biotechnology development. (Freedom to innovate pg 95).

UN-International Year of the Planet Earth (IYPE)

The International Year of Planet Earth is an ambitious program designed to foster outreach and research activities with the single purpose of raising worldwide public and political awareness of the vast (but often under used) potential of earth sciences for improving the quality of life and safeguarding the planet. The aim is encapsulated in the year's subtitle Earth Sciences for Society.

It will run for three successful years 2007, 2008 and 2009 and testify to the Earth's incredible power. More effective use of geo scientific knowledge can save lives and protect property. Such knowledge also enables us satisfy, in a sustainable manner, the growing need for earth's resources by an expanding human population.

The International Year of the Planet Earth was initiated in 2000 by the International Union of Geological Science (IUGS) a non–government organization.

The United Nations Educational, Scientific and Cultural Organization (UNESCO), an international agency, immediately joined the UIGS. The initiative has attracted 12 founding partners that actively support the initiative. The IYPE also enjoys the support of 26 associated partners including all major international geo-scientific and other relevant organizations and growing group of international partners.

The International Year of Planet Earth is a not for profit and non-religious 501 corporation registered in the USA. It is composed of a board, a secretariat and three Advisory bodies.

The Board is responsible for all decision–making on strategic and legal matters and instructs the secretariat. All stakeholders are represented in the Board. These include the two initiators (IUGS and UNESCO) each of 12 founding partners, each group sharing one seat.

The IYPE initiative will further seek to raise the awareness of politicians, decision-makers, the media and the general public on the role and contribution of the Earth sciences in society. The main science themes have been identified for the year, notably, groundwater–towards sustainable use, Hazards–minimizing risks, maximizing awareness, Earth and health–building a safer environment, Climate–the stone tape, Resources–prosperity and sustainability, Mega cities-going deeper building safer, Deep earth–from crust to core, Oceans–abyss of time, Soil–the living skin of the earth, Earth and life-origins of diversity.

Conclusion

The paper on the status and opportunities for investing in Science, Technology and Innovation in Tanzania highlighted in brief the existing situation in the sector.

From the STI status point of view the United Republic of Tanzania is putting a lot of emphasis to make sure that STI contribute to the social economic development of the country. In other words, due to scarcity of resources and what is expected in terms of having an advance STI system that is why the room for attracting investors within and outside, through collaboration comes in.

There are opportunities to make an investment in Tanzania in the STI sector. As pointed above there is a regional initiative to establish a centre of excellence by the name of Nelson Mandela African Institute of Science and Technology. Initially the United Republic of Tanzania, the World Bank has already committed to invest in that centre of excellence. There are other International renowned firms with interest to invest in these initiatives in Arusha once established.

This includes California University, Google etc. Diasporas are also ready to contribute through knowledge building to this institution. With these remarks more rooms are readily available for more investors to invest in this important opportunity for the region and the world at large.

If you are a representative of the industrial sector or of a Foundation you may wish to communicate your ideas about implementation and or sponsoring certain elements of this initiative.

Please note that industries and the Nations of the World need many more scientists to find and sustain ably extract and process earth materials, without which the advance of human society will be seriously impeded.

Acknowledgement

Organizing "International Workshop" is an involving task that requires the collaboration of many players in the course of its preparation. I therefore wish to take this opportunity to thank the organizers first of all for the selection of the 5th International Workshop theme which say's "Enhancement of Change through Science Centres. I can say it has come at the right time, the fact that this time around, the Ministry of Higher Education Science and Technology in Tanzania, is in the process of reviewing the Science, Technology and Innovation System.

May I thank in particular the Organizers and Co- organizers for their tireless efforts and perseverance in investing heavily in all categories so that the workshop become successfully.

Secondly, my sincerely thanks go to the Permanent Secretary of the then Ministry of Higher Education, Science and Technology for appointing me as among the nominee of this important workshop. To me this is an important opportunity in my life time in terms of carrier development and my Ministry especially when it comes to the future and contribution of Science Centres for Social economic development. Again, I wish to express my gratitude to the Acting Director of Commission of Science and Technology as a focal point for accepting my appointment as the only Tanzania nominee.

Lastly but not least, I wish to express my deep gratitude to R&D institutions, namely COSTECH, UDSM, SUA, TIRDO, CAMARITEC etc. who availed the information reflected in this paper.

It will be difficult to mention every one who did its part to ensure that, I become successfully in presenting this paper therefore let me extend my acknowledgement to every one who in a way assisted in making this paper presented as it appear.

References

Archibugi, D., Evangelista, R and Simonetti, R, 1994. "On the Definition and Measurement of Product and Process Innovation" University of Michigan Press, US

Britina, D., Diyamett paper presented in The Status of Science and Technology Indicators in Africa: A Commentary

Calestous, Juma, Serageldin, Ismail. Biotechnology in Africa's development, report of higher level African Panel on Modern Biotechnology.

Ishumi, A.G.M., 1994. 30 Years of Learning: Educational Development in Eastern and Southern Africa from Independence to 1990. Ottawa: International Development Research Centre [IDRC].

Ministry of Science, Technology and Higher Education (MSTHE), 2004. Some Basic Statistics on Higher Learning Institutions in Tanzania, 199/00-2003/2004. Dar es Salaam: MSTHE.

Mwiria, K., 1987. Creativity and Innovativeness among Kenyan Primary School Students. SCR No. 04. Bureau of Educational Research, Kenyatta University and School of Education, McGill University.

Meena, A.S., Pendaeli, J., Chonjo, P. N., O-saki, K.M., 1987. "The Experiences and Problems in Preparing Science Teachers at the University of Dar es Salaam". Paper presented at the Annual Conference of Heads of Secondary Schools in Tanzania, Arusha (Ministry of Education, September 1987).

Mwiria, K., 1987. Creativity and Innovativeness among Kenyan Primary School Students. SCR No. 04. Bureau of Educational Research, Kenyatta University and School of Education, McGill University.

Maliyamkono, T. L and Mason, H, The promise pg.

Nakamura, Takafusa, 1985. Economic Development of Modern Japan. Tokyo: International Society for Educational Information, Inc.

Nkunya, H.H.M., 2003. The State of Science and Technology in Tanzania. Paper presented at workshop on Science and Technology for Africa Renaissance held in Dar es Salaam, Tanzania on June 30th, 2003.

Oyeyinka, B. O, 2004. Buliding Innovative Clusters: The Role of Learning and Local Capabilities, National Meeting, Proceedings of the Regional Conference on Innovation Systems and Innovative Clusters in Africa. Bagamoyo–Tanzania, pp. 153–192.

Owen, Barder, Kremer, Michael and Levine, Ruth, 2005. Making markets for vaccines: Ideas to action: Report of the Centre for Development Working Group (Washington, D.C.: Centre for Global Development), www.cgdev.org (accessed26June 2006)

United Republic of Tanzania (URT), 1989-2006. Basic Education Statistics in Tanzania/Basic Statistics in Education (for various years from 1990-06). Dar es Salaam: Ministry of Education and Culture.

United Republic of Tanzania (URT), 2003. S&T submaster plan 2003-2018.

UNNESCO 2002, Learning through life, Challenges for the twenty-first century.

Abbreviations and Acronyms

- AICAD: African Institute for Capacity Development
- COSTECH: Commission for Science and Technology
- CTBTO: Comprehensive Nuclear Test-Ban Treaty Organisation
- GDP: Gross Domestic Product
- HR: Human Resources
- IAEA: International Atomic Energy Agency
- ICGEB: International Centre for Genetic Engineering and Biotechnology
- ICT: Information Communication Technology
- IPR: Intellectual Property Rights
- MDGs: [United Nations] Millennium Development Goals
- MTAs: Material Transfer Agreements
- NRDA: National Research Development Agenda

- NSGRP: National Strategy for Growth and Reduction of Poverty or [MKUKUTA] [Mkakati wa Kukuza Uchumi na Kupunguza Umaskini Tanzania]
- R&D: Research and Development
- STI: Science, Technology and Innovations
- SMEs: Small and Medium Enterprises
- GOV: Government
- HES: Higher education sector

Chapter 16

Scientific Centres in Cuba: Guarantying Sustainable Development

Lilliam Alvarez Diaz
Cuban Ministry of Science, Technology and Environment, Republic of Cuba
E-mail: lilliam@citma.cu

ABSTRACT

Characterization of the Cuban System for Science and Technology is presented, with special emphasis on the network of 217 important research institutions, distributed along the island, guarantying the development of the 14 Cuban provinces. It will be showing the results of a long term policy focused to the development of the Human resources and infrastructure in a small country without rich natural resources, but strong in education, health and capacity building for Science and Technology. Cuba has been recently catalogued by the UNDP as a country having high human development index, and this fact has been possible only for the permanent political will and investments in S&T and a network of Research institutions working in the main priorities of the country: Food productions, Health, Biotechnology, Renewable Energies, Natural Sciences, Environment, Natural disasters and Climate changes, Basic Sciences, Social Sciences, and Technologies for information and communication. The statistics and some reflections and discussions will show a very particular experience on the crucial importance to invest and develop science in a small country and contribute to the social and economical sustainable development and societal progression in general.

Keywords: *Cuban science, Women in science, S&T indicators, Popularizing science, Science in developing countries.*

Introduction

"One year after the Cuban revolution in 1959, it was decided at the highest political level to link the future of Cuba to the future of science in Cuba. Biotechnology was selected as the focus, reportedly by President Fidel Castro himself. The rationale behind this decision was that, given the limited financial and human resources at the time, S&T had to be development-oriented. Biotechnology was an area of great potential as it could be used for improving health, mainly through the development of new vaccines; increasing crop yield and enhancing animal production; and for earning hard currency through the marketing of biotechnology products". (Ramkissoon[1], 2006).

It is a widely held view that science and technology (S&T) is a necessary tool for development in any society. However, what is lacking in most developing countries is a vision for an effective science policy and most importantly, the political will to implement this policy[1]". "My recent visit to Cuba's research institutes in biotechnology, medicine, agriculture and the basic sciences, and my meetings with the Cuban scientists, left absolutely no doubt in my mind that Cuba's progress in science has not been achieved overnight. Rather, this progress is a result of a vision, planning and implementation involving the political directorate, technocrats and scientists".

Nothing better than an authorized foreign perception, as expressed above, to consider what has happened in the Cuban Science in the last decades. In the paper we would like to express the thesis that there is not possibility to promote changes in a poor o developed countries, without involving all society participating in decisions, changing consume habits, having conscience on the need to take care the environment not giving up to developing. Precisely for that reason the population needs a certain level of general scientific culture, in general aspects of the science, technology and environment and promoting the appropriation of knowledge and scientific results of autochthonous and universal S&T.

First, the Cuban national science system is presented in brief. In section 3 we compare Cuba, as a little and poor country with almost no natural resources, minerals, etc, being only rich in agriculture and human resources mainly.

In section 3, we propose the hypothesis of how to relate classical indicators of S&T with those of human development and in section 4 the importance of incorporate women in the process and actions in the strategy of sustainability.

Finally, in section 5, are shown some practices developed in Cuba to work altogether, starting form the Ministries of Education, Science, Technology and Environment and also the Ministry of Culture (responsible of Medias), involving scientific centres, academies, scientific societies, and the complete scientific community, to increase the general Culture and specifically the Scientific Culture in the whole population.

---

1 At the University of West Indies in Trinidad and Tobago, Prof. Ramkissoon is also Executive-Secretary of Cari science and former President of the Caribbean Academy of Sciences.

Scientific Centres Founded in Almost 5 Decades (1960-2008)

The Table 16.1 shows the sequence of foundation of scientific institutions in the first 20 years:

Table 16.1: Scientific Centres Belonging to the Cuban Academy of Sciences

Year	*Scientific Institution*
1962	Science History Museum "Carlos J. Finlay"
1962	Institute of Geography and Geology
1963	Institute of Scientific and technological Information
1964	Institute of Soils
1964	Institute of Sugar Cane
1964	Institute of Geophysics
1965	Institute of Meteorology
1965	Institute of Biology
1965	Institute of Oceanology
1965	Institute of Literature and Linguistic
1966	Museum for Sciences in Santiago de Cuba
1966	Academy of Sciences in Oriente
1967	Institute of Geology
1969	Institute of Nuclear Physics
1969	Institute of Anthropology
1978	Institute of Meterology
1980	State missions for the Cuban Academy of Sciences

In the 80's measures were then put in place to create new institutions and train a critical mass of scientists, most of them were sent to study and to obtain the PhD in European institutions, not only of East Europe.

13 institutes were created within the Ministry of Health and the National Centre for Scientific Research (CENIC) was established in 1964. CENIC's initial role was to train and prepare scientists to staff a number of research institutes still on the drawing board at the time, of which the most internationally renowned is the Centre for Genetic Engineering and Biotechnology (CIGB), founded in 1986. Nowadays, Cuba has 217 scientific centres and another 63 strong Universities along the island having a population of 11 millions of inhabitant approximately.

The foundations of the Cuban scientific policy continue being: The development of our own human potential and fostering the Cuban social and economical development

The scientific organizations in Cuba, do not work in isolation. They work together in interventions organized by national, sartorial and territorial Programs responding to the priorities of each level and having into account research and development and innovations to solve the concrete problems for sustainable development.

Figure 16.1: National Centre for Scientific Research (CENIC), Established in 1964

The 80´s have been called "the New Era of the Cuban Sciences". An important investment and a set of scientific centres were founded to develop the Biotechnology sector, also with the correct policy to close "the cycle of the R&D", converting in short time the new knowledgements into products. Some of these important institutions are:

- The Centre of Genetic Engineering and Biotechnology
- The Centre for Molecular Immunology
- The National Biopreparations Centre, the main focus of which is the production of Hepatitis B vaccines.
- The *Finlay* Institute, named after national hero Dr Juan Carlos Finlay (1833-1915), who discover that yellow fever virus was transmitted by mosquitoes.
- The National Centre for the Production of Laboratory Animals meets the growing needs of the country's various research centres. The centre produces a variety of animals ranging from mice to monkeys, as well as food for them.
- The National Centre for Neuroscience

Other several institutions, around 36, working altogether to improve the quality of live of the Cuban population and also contributing to the exportation issues with impact in the national economy, (Ramkissoon, 2006).

The idea of presenting data, figures, and indicators of the advances of the Cuban science has the objective to answer the following question: If a developing country increases its capacities in S&T, this fact has a direct repercussion in its human development indicators and in the general cultural of its population?

This critical mass in Science acts as "model of roll" to the youngest generation, improve the public perception of S&T and is guaranty of a future sustainable development?.

S&T for Human Development: Cuba in the Caribbean Region

First of all, let's have a look at the different states of development of countries in the Caribbean: Barbados, Belize, Cuba, Jamaica and Dominican Republic. The Index of Human Development, according the UNDP Inform of 2006, gives a first approximation of the gaps between them:

Table 16.2: Index of Human Development in Some Countries of the Caribbean

Country	*Place in the IDH*
Barbados	31
Cuba	50
T&T	57
Rep.Dominicana	94
Belize	95
Jamaica	104

Notice that Barbados and Cuba classifies as countries with High Human Development, (places from 1 to 63, number 1 is Norway). From the Latin America and the Caribbean Region classifies also Argentina, Chile, Uruguay, Costa Rica, Saint Kits and Nevis, Antigua and Barbuda and Panama.

Figure 16.2: Cuba in the Caribbean

Dominican Republic, Belize and Jamaica classifies with Medium Human Development (places from 64 to 146). Other countries in the Latin America and the Caribbean classifying also in these intermediate places are: Dominica, Brazil, Colombia, Santa Lucia, Venezuela, Peru, Ecuador, Granada, San Vicente, the Grenadines, Suriname, Paraguay, El Salvador, Nicaragua, Bolivia, Honduras, Guatemala, in the same order. Only Haiti, classifies with Low Human Development together with poorest countries of the World, from Sub-Saharan Africa.

Again, let us ask: If a country invests in Education and Science, could improve indicators of quality of life? Can developing countries "dream" or have the ambition to be developed in a medium or long place? Let us see the investments in Education of these countries as per cent of their Total Public expenditure, as is shown in the Figure 16.3.

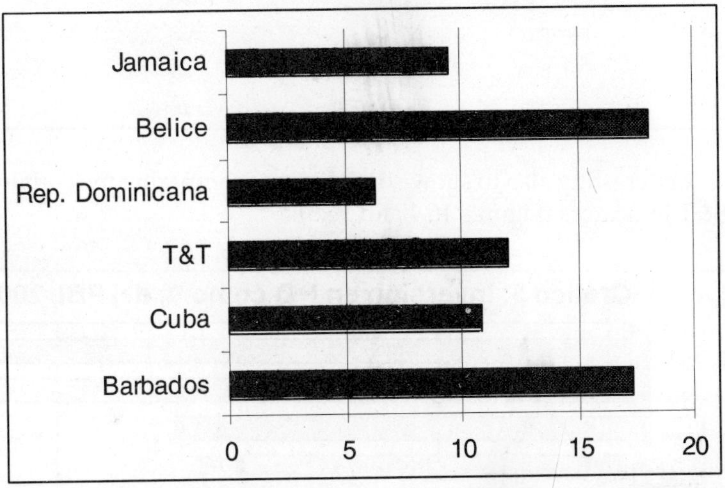

Figure 16.3: Expenses in Education in %

Other indicators, characterizing these countries, are the per cent of Literacy, the Number of children arriving to Grade 5th and per cent of Students in tertiary level in Sciences, Math, Engineering, as shown in the following Table 16.3.

Table 16.3

Country	*Per cent of Literacy*	*Children Arriving to Grade 5th, as Per cent Starting in Grade 1st*	*Per cent of Students in Tertiary Level in Sciences and Engineering*
Barbados	99,4	97	Without data
T&T	96,8	98	Without data
Cuba	95,1	98	Without data
Rep.Dominicana	79,4	59	Without data
Belize	89,1	91	Without data
Jamaica	82,2	90	Without data

It is surprising to notice that such important Indicator for the developed countries and for the UNDP, (that includes students in Science and Engineering at University level), is not computed or reported. However, other countries of the Region report as follows in the Table 16.4.

Table 16.4

Country	*Per cent of Students in Tertiary Level in Sciences, and Engineering*
Argentina	19
Chile	29
Uruguay	Without data
Costa Rica	23
Mexico	33
Panama	21
Brazil	16

It is interesting also to show some Figures comparing the countries, first using two S&T indicators (Figures 16.4 and 16.5).

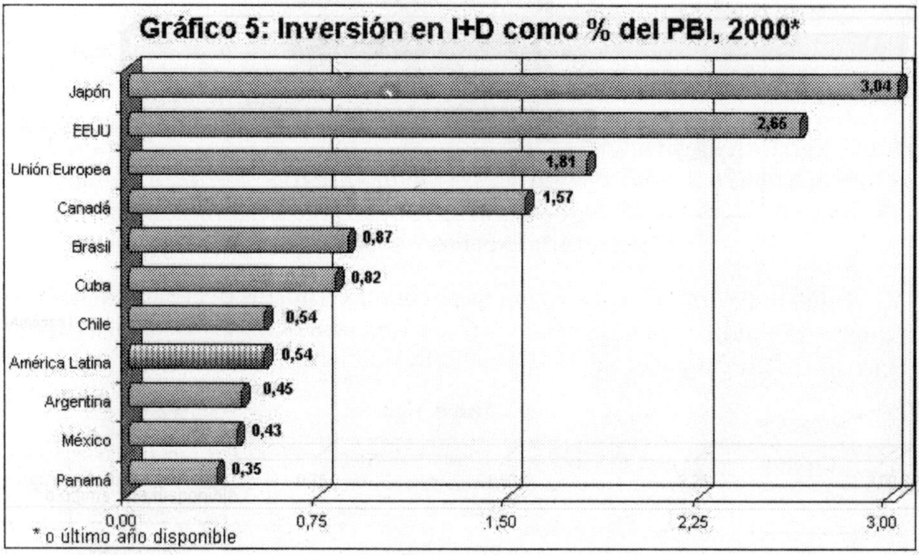

Figure 16.4

Now let us see, one indicator of Human Development, children mortality (Figure 16.6).

And finally see the following graphics showing the relationship between and indicator of S&T with the Human Development Index, as was published by (Albornoz, M., RICYT, 2000).

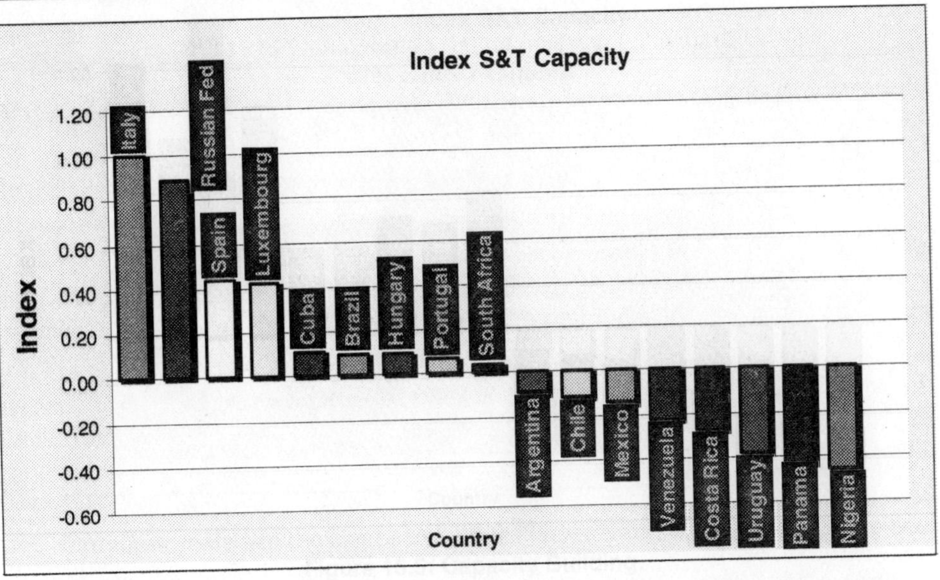

Figure 16.5: Capacity Building

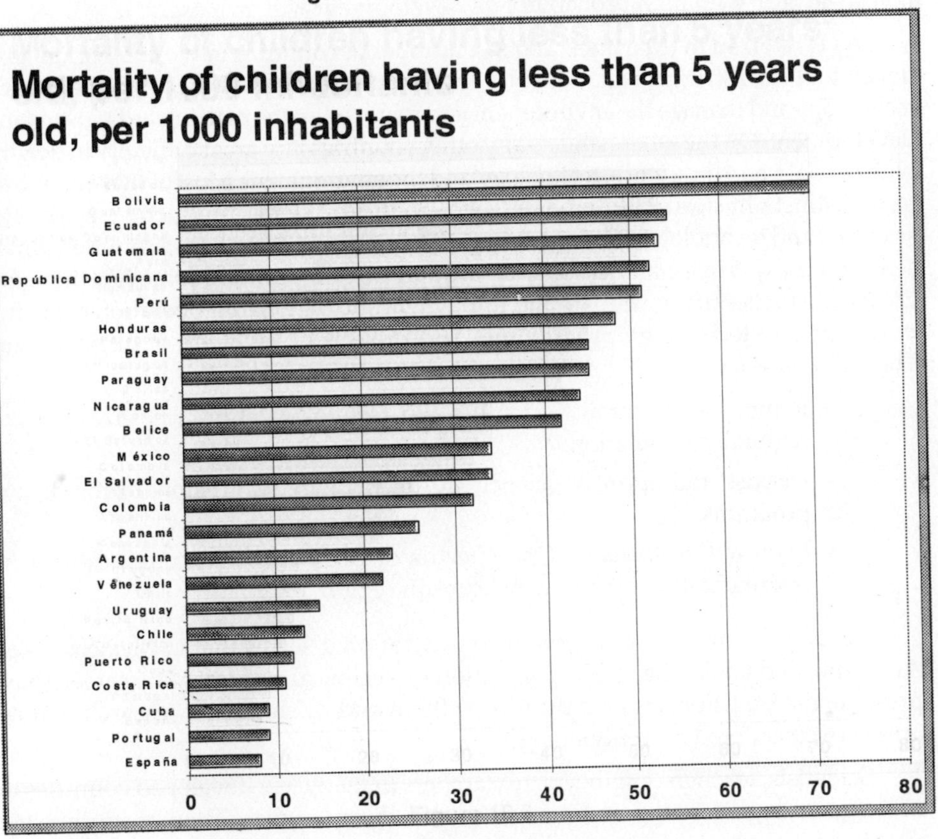

Figure 16.6

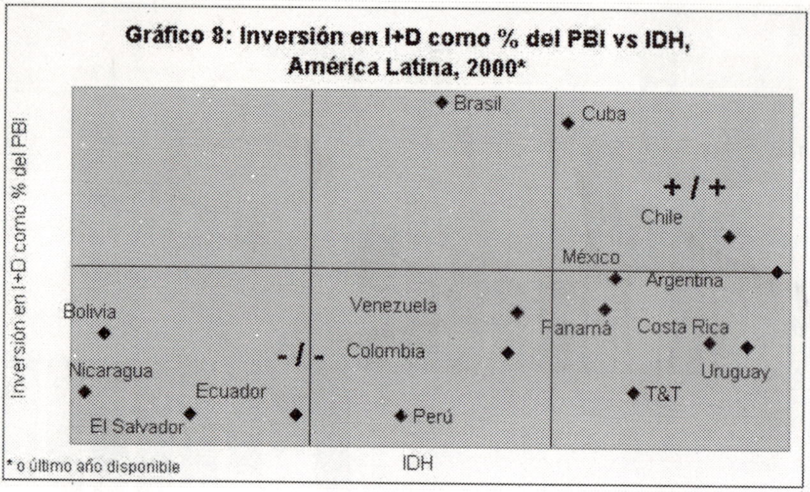

Figure 16.7

This last graphic has the intention to demonstrate that if a developed country invests in S&T with high probability its development will increase.

As is referred in (Martinez, 2005), a number of countries in the sub region have developed infrastructure that enables them to meet the challenges of science and technology and manage the environment and might be on the path towards sustainable development. On the other hand, many other countries face great difficulty in dealing with crucial issues in science, technology and the environment. Most of those countries have distinct physical environments, socio-political systems, productive sectors and scientific and technological infrastructure. They nevertheless share a common aim to share their experience and co-operate around priority issues. Moreover, there is a great need to identify, formulate and implement sub regional co-operation measures in science and technology programmes and projects in key areas of development. There is a need to:

- Identify and promote scientific and technological priority areas for sub regional co-operation
- Discuss and approve science and technology sub regional projects and programs
- Review the status and prospectus of R&D and science education and popularization in the Caribbean sub region

Beyond the figures, as a general consideration it seems that in Mexico, Brazil, Argentina and Chile, the public perception on technical and scientific careers have prestige, the S&T Policies give priority to them and teaching and research in these areas are considered a livelihood.

Scientists, teachers, engineers, researchers in many less developed countries do not earn enough to support themselves and their families. Then, it implies two situations: young people do not go to scientific or technical careers and if go, the next step is to immigrate to a developed country to continue their professional lives, (Zink, 2006).

Vital Necessity: To Incorporate Women in the S&T Systems

In this section we present country-level data on the participation of Cuban women in different scientific fields, having scientific degrees, occupying decision positions and comparisons with other countries in Latin America.

The social analysis of the not-so-obvious problems has been done by the author and other scientists in the Region, (Alvarez, 2001, 2002, 2003) and have been done from the perspective of women doing basic sciences, in particular women in Physics.

In Cuba, in spite of having a well established political Will for more than 40 years, that opened opportunities for everybody, opportunities to all the talent youth, including girls; still are present stereotypes, prejudices in our culture, and still we find professional bias, troubles in getting positions of leaderships in some sectors, lack of transparency in some decision-making levels, etc.

Only few years ago the basic sciences, as Physics and Mathematics, have been approved as national priorities in the National Plan for Sciences. Of course the previous programmatic direction affected female and male physicists alike, regarding resources allocation and project supports, but had special negative impact for women.

We discuss the importance of having women in leadership positions and how they can promote and serve as a pattern role model for the next generations. Also it is important to put in first place the support of the family, including parents and husbands as a factor of the success of many women physicists especially in developing countries.

The necessity of mobilization, to adopt recommendations on access to scientific careers, science policies, networking, and other important topics are crucial to shape the future of women in sciences.

In concluding, some directions for work in the near future, for collaboration, in order to strengthen efforts at ensuring the full participation of women and girls in all aspects of science and technology in Cuba. In addition, recommendations have been made on ways of overcoming existing gender stereotypes among scientists, technologists, policy-makers and the community at large in our Latin-American and Caribbean countries.

It is well known that political will by a government is crucial in the struggle to incorporate women in all the social spheres, and in particular in the Science and Technology domains. Since the very beginning of its triumph, the Cuban Revolution opened equal opportunities for males and females, and many concrete actions were taken in the early 1960s. These actions included:

- The Campaign against illiteracy in 1961, where 59 per cent of the young, volunteer teachers were women and 55 per cent of the people learning to read and write were women;
- Founding of the Cuban Women Federation (FMC), an organization that played a decisive role in recovering the dignity of Cuban women, by

involving them in the same activities as men, with complete recourse to the labor laws. The FMC founded the Children Garden and facilitated the inclusion of lunch in many new schools. These were definite steps in adding the contributions of women to the construction of a new society.

- The Pregnancy Law and the Family Code gave official character to gender equality, in terms of economic and political opportunities and women's access to decision-making levels.
- The new Pregnancy Law (2001) guarantees one year full salary to mothers.

As a consequence of this political will, Cuba today proudly shows impressive statistics for the participation of women in all aspects of society, as shown next:

Table 16.5: Female Participation in Cuba's Social Sector

Area	Per cent Technical force	Per cent Leadership
Health	79.9	35.8
Education	70.2	48.6
Tourism	52.0	25.0
Science and Technology	**66.6**	26.1
Sugar agro-industry	33.9	10.5
Agriculture	32.6	11.6
Cooperative sector	28.2	14.4

Other important facts to show is the access of Cuban women the postgraduate studies. Comparing with the total human resources working in Cuban Universities and Scientific institutions the percentages are, as average in the five last years, as follows in Table 16.6.

Table 16.6: Cuban Women having Postgraduate Studies

Sl.No.	Qualification	Per cent of Women
1.	Decent positions	46,43
2.	Researching category	40,53
3.	PhD	24,16

In Figure 16.8, we show the per cent of men and women working in Research and Development activities in different countries of the Region. The source is the Network for S&T indicators, RICYT, 2000. We can see that Uruguay and Cuba have better positions.

Despite, the above statistical data, it is possible to observe in Cuba, traditional and historical vestiges of gender bias. For example, women continue to be promoted at a slower rate than men. Also the 'glass ceiling' phenomenon is experienced by some women in all areas of work in the Cuban society.

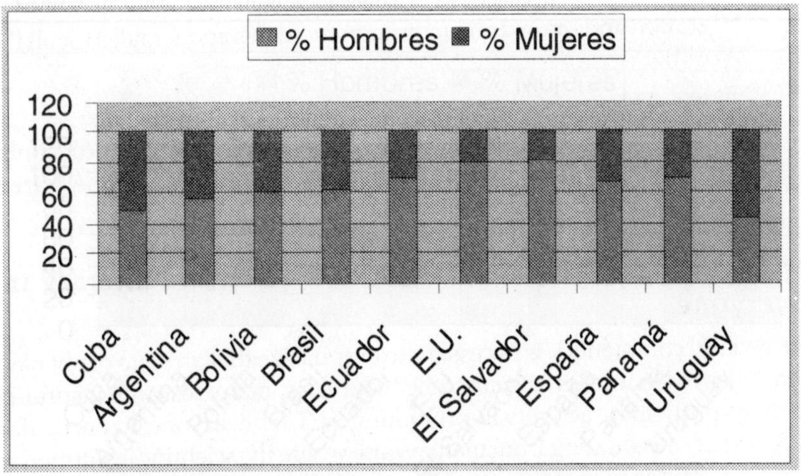

Figure 16.8: Percentage of Men and Women Working in Research and Development Activities (Hombres–Male; Mujeres–Female)

Of course, compared with Sub-Saharan Africa, some Asian regions and the Caribbean, Cuba is in a better position with respect to the involvement of women in science and technology. This, however, is not true when the comparison is made with Southern countries (as Brazil, Uruguay, Chile and Argentina), which all have similar indices for access of women to scientific careers and studies at the postgraduate level.

The African and Asian Diaspora has to be taken into account because it is a reality in Caribbean and Latin-American countries. Of necessity, the historical and cultural basis and the local inequities prevailing among the different races in the region have to be considered. Many women in science and technology in the Caribbean feel pressured and judged not only for sex but only for race. Because of that, they have continuously had to 'prove themselves' and can lead them to develop extraordinarily high standards, as a perquisite for staying in their scientific fields.

During fruitful debates in different international conferences on women in sciences and exchanging much different point of views the author has identified the following problems which are common to women pursuing careers in science and technology:

- Leadership Problems.
- Under-Representation of Women at Decision-Making Levels.
- Occupational Segregation.
- Tendency to Medical/Health Sciences and Education–which are Traditional Vestiges.
- Under Representation of Women in Basic Sciences as Mathematics, Physics, Astronomy (except Chemistry).

- Migration and brain drain (common for males and females for all the Third World).
- Stereotypes.
- Gender Bias in Text Books for all Levels and in Scientific Vocabulary and Scientific Literature. (Sirju-Charran, 1996; Broomes, 2001; Alvarez, Sirju-Charran, Broomes, Doelhalasori, 2003).

Some Practices for Popularization and Scientific Literacy in S&T for Everyone

As general comments it is necessary to emphasize that, in spite of not having yet in Cuba, Science Centres as in many other countries, many research institutions and Universities participate actively in promoting S&T for everyone. In particular, there are many experiences dating from many years, where the scientific institution receives every Friday, children, parents, grandparents from the neighborhoods where the centre is placed. They organize visit to the centre, seminars, contests, recreational activities, etc.

Examples of these are not only the most obvious centres designed for scientific recreation as Aquariums, Zoos or Botanic Gardens, but also centres as the Institute of Pharmaceutical Chemistry or the Institute of Tropical diseases. This last one, specialized in *Dengue*, every week has also a TV program explaining how to eliminate mosquitoes from our houses.

There exist also wonderful experiences in the organization of TV courses, named "University for Everyone", where the most relevant Cuban scientists, leaders and also young researchers, teach courses by television using modern techniques of social communication of Sciences, with topic going from Introduction to Biotechnology, The Sea and its Resources, Global Climate Changes, The Geological nature of Cuba, to Chess for everybody, Languages (English, French, Portuguese), etc. The surppresive increasing of the television audience derived to the decision of opening two Educative Channels, to promote scientific short materials, and also to support every primary and secondary schools with TV classes, helping teachers and students.

In Cuba, in every classroom there is a TV and a Video and each teacher has, as media, only 20 students. If the school is placed in the mountains or in isolated places, sometimes with schools having only five children, there, it has been installed photovoltaic panels to use solar energy for the TV and video of that school. And at night, with the batteries, the TV is used for the neighbors and the families living there.

The Ministry of Education of Cuba, has already established and also since many years ago, activities of Teaching Sciences in non-formal ways. National contests, scientific societies for Students, the so called "Círculos de interés" for different disciplines, the National Olympiads to choose the students for participation in the International ones, in Physics, in Mathematics, in Informatics with prizes and excellent results.

By other hands, the Universities also have many activities, especially directed to young people, motivating and given vocational orientation to students willing pursue scientific careers, especially those studying in the so called Vocational College of Exact Sciences. Once a year, every University in Cuba organizes the "Door Open Day", where Professors and researchers receive girls and boys in their laboratories and classrooms. Also, there is a practice since at least 30 years of the development of Scientific Student Journeys, where the students from third, fourth or fifth year present, discussed and participate in a competition of their researching projects in their careers.

Every University also has its own Department of "Extension", to organize cultural activities related to artistic culture, but in the last years are joining efforts to promote also the scientific culture.

The scientific societies also have, as its normal duties, to organize activities for popularization its respective disciplines and all of them are supervised and helped by the Academicians, full members of the Academy of Sciences and its Executive Board. It is notorious that young researchers are eligible to be Academician for two or three periods (8 years).

Let us expose some recent projects involving the Cuban scientific community in general.

Project "Seeding Sciences for the Future"

In the current world, economic and social development requires, among other important actions, the development of mass and continuous education programs, promotion of culture and, at the same time, achieves appropriation of Science and Technology, so that it provides for the creation of value-added processes and knowledge transmission and diffusion.

Cuba is a country in which science and technology capacities have been developed under very difficult conditions. However, these capacities can be exploited even better, not only for the benefit of the country, but also of the countries of the area, especially of Central America and the Caribbean.

Cuba has created capacities, mainly human resources, able to promote high-quality scientific and technological education, through formal and non-formal ways, integrating science and technology discoveries and inserting them in every day life and community problems. This allows creating an active and creative position towards science, technology and environmental protection. However, these capacities are not being properly exploited.

The proposed project has as institutional framework the Division of Sciences of the Ministry of Science, Technology and Environment, (CITMA) that has, among its missions, fostering socialization of science achievements, their introduction in the educational system and their efficient divulgation to elevate the scientific culture of the country. The Ministry of Higher Education (MES), the Ministry of Education (MINED) and the Municipal governments collaborate with the project.

The project consists of a group of actions of science promoting and popularization that will contribute to the construction of a knowledge society, (knowledgement for everyone). The municipal structure of CITMA specialists are used for this objective, multiplying its current capacities so that it contributes more effectively to promote, foster, divulge and socialize the results of science, technology and environment protection in each Cuban municipality. This would represent a considerable force to achieve a greater social appropriation of knowledge, resulting, in turn, in a greater impact of Science and Technology on the community, which has a hard pressing need to raise its scientific and technological education to assimilate the results of science itself.

The results of this experience can be applied forwards in other countries of the area, as an alternative that Cuba offers to solve the knowledge appropriation problem that strikes our poor countries.

It is sought to create a network which allows searching and interchanging information and favors the design and implementation of interactive activities in community institutions, motivating and raising the interest of all the society, specially children, adolescents, young adults and women. This would facilitate promotion of a higher scientific, environmental and informational culture, and achieve a new perception of Science and Technology in the municipal setting, as a way to solve the sustainable development problems and improve the population's living standards.

Parallel activities to strengthen and increase the incorporation of students of the primary and secondary education to Scientific and Technical clubs are carried out.

The planned activities permit to "seed" the necessary calling towards science careers and originate an active and creative position towards science and technology. Implementation of these activities, that promote scientific education of all the population, facilitate that future generations acquire a greater responsibility in the search of alternative ways of scientific and technological development and feel motivated to control the impact that scientific and technological activities have on daily life.

Main Objective of the Project

To contribute to social knowledge appropriation by means of the promotion of scientific and technological education at the community level, using the potentialities created by CITMA in the municipalities.

Specific Objectives

- To strengthen incorporation of scientific and technological knowledge in the national culture by means of the implementation of interactive activities of scientific divulging in community institutions, contributing at the same time, to motivate creativity.
- To mobilize a group of community institutions in order to develop interactive activities and exhibitions that provide for collective experimentation of knowledge and offer diverse ways towards appropriation of scientific, technological and environmental knowledge.
- Strengthen scientific-technical clubs in municipal elementary and secondary schools and increase the incorporation of children, adolescents and young adults.

- Develop a knowledge management network for the community, which would be known as "Seeding science for the future network", which permits to generate and to share knowledge, by means of the connection to Internet of the 168 CITMA municipal specialists.
- Improve the current infrastructure of the 168 CITMA municipal specialists so that they can contribute more effectively to promote, foster, divulge and socialize the results of Science, Technology and protection of environment in each Cuban municipality.
- Benefit other countries of Central America and the Caribbean providing access to the experiences and results attained.

Annual Journeys to Increase the Scientific Culture in the Cuban Population

Identified Issue

We use to say in our diary speeches, that Cuban population has acquired a high political culture and also high culture in sport issues, but that is not the situation in terms of scientific culture. In the Cuban Ministry of Science, Technology and Environment, specifically in the Direction of Sciences, we have a medium and long term Program with several actions to improve this situation. This objective is involved in a more general Governmental plan called the Battle for a General Culture of the Cuban people.

Engage the scientists in social communication activities are a rather complicated job. First, because, sometimes they do not have the social commitment to spend some of their valuable time going to record some radio or TV programs or going through the all island to share, in a rigorous and simple language, their scientific background. In another hand, when you find someone with the willing to enface the job, usually the researcher does not have communication capacity.

Every year, in the Cuban Ministry we organize a national meeting that we call *Journey for a Scientific Culture,* that has become a very important moment when many professionals, journalists, decision makers, public in general have the possibilities to hear the most eminent Cuban scientists talking about the more relevant and hot topics of the universal science and also about their own researching.

Science Communication Objective

- To promote and popularize the developments of Science and new Technologies.
- To improve and update the knowledge of specialists on top scientific topics of special interest.
- To popularize the latest advances of science
- To contribute to the development of a Scientific Culture among specialists and executives in Science and Technology Systems in the Region

Target Audience

The *Journeys for a Scientific Culture* are intended for executives, officials of the Systems of Science and Technological Innovation in Latin America, specialists, professors, teachers, students, journalists and media professionals, and all those scientific workers interested in increasing their Culture on the cutting edge challenges of Science in the XXI Century.

It is a relevant fact, that we include in our audience the workers of science, technology and environment of every municipality of our country, in such a way that having they all the materials, CDs, documentations, films, etc, they have the possibility to multiply the experience to rural communities, schools, neighborhoods, etc.

The Science Communication Process

The Journeys are held in the halls of our Capitol Building in Havana. They have excellent structure and infrastructure to develop the meetings. Every year the three day event is dedicated to relevant thematic lines, for example, Genetically modified organisms, Stem Cells, last results of the Cuban sciences in Vaccines or Biotechnology, other edition have devoted to Science and Technology for environment or Sustainable Development, Global Climate changes and specifically this year is on behalf of the "World Year of Physics", with topics about the Big-Bang or the origin of the Universe, Einstein in the dairy life, Women in Basic Sciences, etc.

In the 2003 edition of the Journeys we have the visit of a Nobel Prize in Physics. L. Lederman, with a wonderful Conference on Physics Teaching.

In every edition the most relevant men and women in sciences in Cuba and Latin America are invited to teach Special Conferences, to show experiences or films, and all the activities are organized as a *Show Talk* style. In the process of the *Journeys* we include the modality called *Café Scientifique* to promote open discussions about the public perception on Science and Technology.

We combine the popularization activities with music, in general classic music, promoting that Science is part of the National Culture and a complement of the spiritual world of the citizens.

Results Achieved

The impact of the *Journeys* has been important in the scientific *panorama* in Cuba. As long as we know, there were isolated experiences in different places and moments of the scientific development in the last four decades, but not a permanent national meeting having these characteristics.

Every year already the scientific community, young people, students and teachers, journalists dedicated to scientific journalism, administrative workers of the S&T systems, wait every year the days of the Journeys (usually every April).

We modestly think that we have achieved three main results:

- To engage every year Cuban scientists and Academicians in the communications activities

- To stimulate the interest of knowing more about Science and Technology of a wider group of the Cuban people
- To contribute to the appropriation of basic concepts on S&T of, every year, more number of citizens, increasing their scientific culture.

Evaluation/Assessment

The Cuban Program to increase the Scientific Culture is part of the Objectives of the Cuban Ministry of Science, Technology and Environment for the next years. So, every six month this Program is evaluated, and the *Journeys*, as part of the Program, are discussed in terms of organization, support, thematic lines, etc. In all the occasions the *Journeys* have been evaluated as a successful experience, very necessary for the direction of Sciences mission and functions.

TV Program: *"Pasaje a lo Desconocido"*, Passage to the Unknown, (Alvarez, 2005)

Science Popularization in general, but in special on TV, is, in a Third World Country a difficult job, in order to produce attractive materials for a broad audience. Otherwise, the main stream on TV production has excellent materials and sources to get immerse a broad public in the cutting edge scientific research and its development.

A way to show the advances of the national scientific community in fields like Astronomy, Astrophysics, and Cosmology, applicable to other sciences also, connecting them in friendly relation with a majority of the people, is to combine the wisdom and knowledge of the local scientists together with the most spectacular TV production of the first world countries, commenting, analyzing and conveying the hard science into the public debate of the common citizens.

Science Communication Objective

Convey cutting edge science to the general public of a third world nation, with limited resources to produce TV imaginary and filmography, highlighting and making public the development, knowledge and wisdom of the local scientists.

Target Audience

All the population nation wide, from the most humble farmer to the higher specialist in fields other than the treated topic of each presentation, policy makers, children, young persons and elder people.

The Science Communication Process

A confront interview, where a leading journalist and commentator of the media which assumes the role of the common people generally non familiarized with the theme, deploy their imaginary and questions on the treated topic, interpellates an experimented national scientist which has to sort the difficulties of the scientific terminology and custom language in using a plain explanation to be accepted by the public, keeping a high rate of audition attentive.

Results Achieved

A high rate of public attentive, placing the presentation on the highest levels of the scientific programming, not very far from the most popular programs of the general entertainment schedule and reaching the pursued objective to involve about three or four million spectators of the national population in the scientific controversy.

Evaluation/Assessment

Index rate of the TV program that convey the experience oscillating in the last five years no less than 22 up to 32. Source: Centro de Investigaciones Sociales (CIS) del Institute Cubano de Radio y Televisión (ICRT).

Conclusions

There are many other activities enhancing change through governmental institutions in Cuba. Many experiences for popularizing ICTs, called the "Joven Clubs", the National Forum for S&T, that is a massive movement involving thousands of Cubans every two years, including housewives, retired workers, farmers, are already a reality in Cuba. By other hands, the social scientists have to be involved since the very beginning, to measure the changes, to measure the social, economical and cultural impacts of these actions and projects.

Of course that still there are many obstacles, barriers, lack of comprehension, lack of conviction of importance of S&T, lack of financial support, lack of integration, but, the Cuban experiences show that having, first a political and governmental will and second a scientific community with a very strong moral and ethical commitment with its country and its citizens, is possible to do a lot, looking for a better future and a better world using the Scientific advances.

Acknowledgements

L. Alvarez is deeply grateful for the support of the Centre for Science and Technology of Non-Aligned and other developing countries, and also of the South African authorities that made possible to write this paper and to participate in this 5th International Workshop.

References

Alvarez, Lilliam, 2001. Women in Sciences in Cuba Today. Invited Lecture in the Conference on Key Issues facing women in Science in the Caribbean held in Trinidad and Tobago (29th April to 2nd May, 2001).

Alvarez L., Aurora Perez Martinez and Margarita Cobas, 2002. Conference of Women in Physics, Paris, published by AIP Proceedings, Vol. 628, pp. 151-152.

Alvarez, Lilliam, Sirju-Charran, Grace, Broomes, Veronica, Doelhalasori, Rinia, 2003. Preliminary remark on the situation of women in science and technology in the Caribbean, Conference of the Caribbean Academy of Sciences.

Alvarez, Oscar, 2005. Cuban-Japan ODA project for a modern Planetary in Havana.

Alvarez, L. and Olivé, E., 2005. Seeding science for the future, presented as UNESCO project, Havana.

Broomes, Veronica, 2001. A Journey of Challenges–Women in Science and Technology in Guyana. Conference on Key Issues Facing Women in Science in the Caribbean held in Trinidad (29th April to 2nd May, 2001).

Martínez, E., 2005. Call to the 1st UNESCO Conference for Science, Technology and Sustainability in the Caribbean, held in Havana.

Ramkissoon, Harold, 2006. Cuban science: a jewel in the Caribbean crown, published A World of Science, UNESCO Natural Sciences Newsletter,. Vol. 4, No. 3.

Sirju-Charran, Grace. 1996. Women in Science: the Trinidad and Tobago Experience, the Role of Women in the Development of Science and Technology in the Third World.

Marcia Barbosa, 2003. Women in Sciences: The future of Physics depends on Women, (private communication).

Watkins, Kevin, Director y Redactor, 2006. UNDP Inform on Human Development,

Zink E., 2006. Summary of IFS Impact Studies, Report No. 6, pp. 10-11.

Chapter 17

The Rise of a Science and Society Culture in Turkey

Ilay Celik* and F. Kubra Gokdemir**

The Scientific and Technological Research Council of Turkey (TUBITAK), Ankara, Turkey

*E-mail: *ilay.celik@tubitak.gov.tr, **kubra.gokdemir@tubitak.gov.tr*

ABSTRACT

The Scientific and Technological Research Council of Turkey (TUBITAK)'s Science and Society Division aims to contribute towards the adoption of a scientific approach in society, development and dissemination of scientific culture, and supporting policy making efforts to become a science-oriented society. TUBITAK commenced periodical calls to support activities to popularize science and technology among public. TUBITAK diversified its science and society project range by designing novel projects. The total budget of Science and Society Division extends up to $26 million in 2008. This paper describes the TUBITAK Summer Science Camp and Science & Society Project Support Programme in detail. TUBITAK Summer Science Camp is a novel model project. This project hosts 11 year-old kids from a large scope of geographical locations in Turkey. The workshops of the camp involve applied, interactive, fun, and participant-focused activities. The objective of the camp is to help participants associate science with daily life and adopt scientific thinking. As an efficient way of disseminating similar activities, TUBITAK decided to motivate Universities and Governmental Institutions for designing and applying similar projects and started a programme for supporting science and society projects. Two project calls have been opened so far: Science Camps/Schools and Nature Education. In 2007, the total number of the projects funded is 27 out of 83 candidate projects presented by Universities and Governmental Institutions. The objectives of these projects are to publicise and promote scientific concepts and innovations in society to develop a nationwide awareness of scientific thinking and show that science is interesting, fun, and attractive. Supported projects have reached up to 3500 people of different age groups and from various locations in Turkey. A science and society culture, which will certainly contribute to development of Turkey, has already begun to rise.

Keywords: *Science and society, Science camp, Scientific culture, Support programme, Popular science.*

Introduction

In 2004, TUBITAK conducted a research on Science Communication in Turkey. This research demonstrated the present attitudes towards science, scientists and TUBITAK among 15-24 year-old Turkish citizens. The youth population between the age of 15-24 constitutes 20 per cent of total population of Turkey. In other words 14 million people in Turkey can be considered as young.

According to the results of the research, young people in Turkey are not against science and scientific thinking but, they perceive science as a concept distant to "normal people". They think doing science is beyond their capacities and capabilities. They also find dealing with science boring, difficult and tiring. They perceive scientists as unsociable and insensitive to their surroundings.

This and other results of the research motivated TUBITAK to take actions in the field of Science and Society and this area was taken into priority issues. TUBITAK established the Science and Society Department in 2005 and started new projects to flourish science and society culture in Turkey. In fact, TUBITAK has worked for popularizing science for years, through its popular science publications and popular science activities. In recent years, TUBITAK diversified its science and society activities and established the project support programme in order to further disseminate scientific culture. Previous popular science activities of TUBITAK included Invention Festival to improve and reveal children's capability of innovative ideas, Sky-Observation Festival to gather amateur astronomers and professional astronomers, Formula G and Hidromobil-07 events to challenge university students to produce and race alternative energy powered cars, Underwater Science Camp to improve the participants' underwater research abilities. Most recently TUBITAK Summer Science Camp was added to this range of activities. Other long-term science and society projects of TUBITAK are production of scientific toys and observation kits and the establishment of a vast, world-class Science Centre in Istanbul.

The Science and Society Project Support Programme was started in 2007 in order to fund science and society projects designed by governmental institutions and universities. Therefore, this programme provides opportunity for more people to benefit from popular science activities and it serves well to out-reach purposes. This paper describes one of the model projects of TUBITAK, the TUBITAK Summer Science Camp and the Science and Society Project Support Programme in detail.

TUBITAK Summer Science Camp

Among all its target groups, TUBITAK gives priority to pre-school and primary school kids depending on the well-known fact that it is much easier to influence people's attitudes and behaviours at early ages. TUBITAK has previously been organising activities addressing primary school kids. Invention Festival has hosted primary school kids of 7 years old. Kids send their inventions prior to the festival and all inventions are exhibited and evaluated for prizes during the festival. Besides, there are invention-related activities for kids of different age groups. The Sky Observation Festival, hosts participants from all ages and primary school kids constitute a high proportion. In 2006, when TUBITAK decided to diversify its science and society activities, TUBITAK started the Summer Science Camp for primary school kids. Thus TUBITAK Summer Science Camp was established as novel model project addressing primary school kids.

The first pilot project was designed in a few months. Since it would be a pilot project, participation was restricted with a single age group. Participants would be primary school kids who completed 5th class and would start 6th class. Since the quota had to be limited, another criterion was applied: being a subscriber of TUBITAK's Science Kid magazine. Therefore applications are accepted from Science Kid magazine subscribers who just complete 5th class of primary school. These criteria allow us to make a restriction depending on an objective basis. The participants are determined by a lottery by notary among applications.

The number of participants for a single group is determined as 32. It is a feasible number considering the age and the programme. Participants are accepted in equal numbers for two genders. In order to encourage participation from developing regions, total quota is divided between developing regions and other regions according to corresponding county numbers.

When the content was being determined, special effort was made to keep the subjects as variable as possible. Since it would be a "science camp", some general and basic subjects should be included.

The philosophy behind designing such a camp, as shortly expressed in the motto "Science is everywhere!", is to make people aware of that science is not something disconnected from life, that scientific facts do exist in every area of daily life and that doing science is not something boring, on the contrary can be great fun. Therefore the primary objective would not be conveying a lot of information through the activities but rather showing the participants the enjoyable and charming face of science and thus trigger their curiosity and desire for inquiry. Conveying information would not be the aim but a tool. Above all, all activities of the camp should somehow involve fun and game.

The camp is done in an institute of TUBITAK, Turkish Institute For Industrial Management (TUSSIDE), which possesses complete facilities for the camp. It has a high standard hotel, many halls, and open areas. Two participants share a room during the camp. One camp period continues for 6 days.

The staff of the camp consists of 3-4 TUBITAK employee and about 15 guides, who are volunteer university students. There is a division of labour for a more effective operation. Each of 5 guides is responsible for a participant group of 6-8. They are primarily responsible for assisting and guiding their group during their stay in the camp and caring for them all the time. Other guides are primarily responsible for preparation and execution of workshops and activities during the camp. However, all guides are responsible for both participants and activities on the whole.

The guides are chosen from applications prior to the camp. The camp is announced in a few universities and applications are collected. After a pre-selection from application forms a meeting is held with the applicants. In this meeting the applicants are asked to answer some questions that provide information about how suitable the applicant is for this camp. In addition to what the applicant tells, the way he/she expresses himself/herself and the way he/she behaves in a group provide valuable information in selecting the guides. Some criteria for selection of guides are good communication skills, previous experience with children or temperament that makes one potentially a good science communicator.

TUBITAK's previous experience of science society activities contributed much to the content of the camp. Astronomy workshop was designed depending on the experiences from previous Sky Observation Festivals, which was then 8-year-old; similarly Invention Workshop benefitted much from experiences from Invention Festivals.

The programme of the camp consists of half-day workshops. Some of these workshops are done in TUSSIDE and some are in different research institutes of TUBITAK. Most of the remaining time is reserved as free time for participants. In free hours, participants are allowed to do whatever they would like to. However, many alternative activities are kept ready for their choice. These activities include, open air games and board games and thematic activities such as photography or capoeira. Participants arrive at camp in an afternoon and after they settle down, they meet each other and the camp staff through meeting activities.

The first day of the camp starts with the workshop "Science in the Kitchen". The workshop was designed in order to demonstrate the connections of chemistry science with daily life. In this workshop participants make simple chemistry experiments using daily kitchen material and equipment. The workshop is directed by a guide and each of six subgroups is also assisted by a guide. The workshop is done in a kitchen scene. At the beginning of the workshop participants make pastries and then do the experiments. The experiments are associated with the process of making pastries. The second workshop of the first day is done in the Chemistry and Environment Institute of TUBITAK. In this workshop participants do experiments in a real research environment with researchers. One of the experiments is related to one of those done in the Science in the Kitchen workshop, therefore it is possible to establish an association in between. They find the opportunity to meet and talk to scientists and to learn about their lives. The first evening, participants see an enjoyable film, which is somehow related to science.

The second day starts with the "Nature Workshop". This workshop is done in the open air. TUSSÝDE is surrounded by natural areas. Participants are guided by the workshop executive and assisted by other guides. Participants join activities that allow them to make nature observations and to learn about biodiversity. They take water samples from the pond near TUSSIDE. In the afternoon, a workshop takes place in TUBITAK's Institute for Genetic Engineering and Biotechnology. In this workshop participants examine their water samples taken from the pond and then they make DNA isolation from fruits. After this workshop the group returns to TUSSIDE and participants play open air games about nature. In the evening as the conclusion activity of nature workshop, participants make pictures about their nature observations on very large cloths, using different kinds of dyes and employing different parts of their bodies.

The first workshop of the third day is Archaeology workshop. The workshop starts with a brief presentation and an activity called "Archaeology in the Dustbin", which introduces the methodology and logic of archaeology. Later participants join an excavation in an artificial excavation area. They use similar techniques to those archaeologist do. In the end of the workshop participants exhibit their foundlings in a museum scene and they introduce them to visitors. The second workshop of the day is Astronomy workshop. In this workshop participants listen to a presentation that gives basic information about planets, stars, the Solar System and constellations through attractive visual material. And then participants play a game in which they form the Solar System using balloons and tissue paper. After this activity, participants join an activity in which they create their own constellations and create stories for them. In the evening they make sky observations both with naked eye and using a telescope.

The workshop "Detectives on Duty" is the first activity of the fourth day. In this workshop participants try to solve a robbery event using scientific methods and scientific way of thinking.

The game starts with a robbery event at a museum. The participants then become detectives and they solve different parts of the puzzle. The fourth day continues with the Invention workshop. This workshop introduces the approach and way of thinking behind inventions and gives the idea that everybody can make inventions. Participants make a brainstorm about inventing with guidance of the workshop executive and make their own definition of invention. Then they invent super gloves. They make a mechanism that prevents a thrown egg from breaking and they construct platforms which can carry tens books, from eight A4 papers.

The last day of the camp starts with Robot workshop. In this workshop participants programme robots using an interface that enable them to give instructions to robots without using programming languages. The workshop starts with an enjoyable and visually rich presentation, which gives brief information about what a robot is and is not, and about the basic features of robots. Then participants work in groups to execute a mission given to their robot. The last workshop of the camp takes place in TUBITAK's National Metrology Institute. Participants join experimental activities about electromagnetism, pulling and hardness tests and photo-acoustics with the guidance of researchers.

There is a party the last night of the camp. There is food and music in the party. It is a party just for fun and celebration. The friendships and the friendly atmosphere of the camp are reinforced in this party. Thus all good memories of the camp are further strengthened.

The last morning there is an exhibition in which any physical outcomes are exhibited to parents and participants.

TUBITAK Summer Science Camp hosted 60 participants in 2006 and 128 in 2007. It is planned to continue the project. The content and the organization of the camp are improved every year to set better standards for science camps to be designed by other institutions.

Science and Society Support Programme

TUBITAK commenced periodical project calls to support activities that aim to popularize science and technology among public. TUBITAK has published the initial project calls entitled Nature Education and Science Camps and Schools in March 2007.

Firstly, the Nature Education will be explained in the light of past experiences on Nature Education of TUBITAK. The general aim of Nature Education is to provide a multidisciplinary approach that develops the individuals' skills to observe and evaluate the mechanisms of nature from a scientific perspective and adopt a new ecological vision of nature. TUBITAK had conducted the projects of Nature Education in cooperation with academicians from various prestigious universities in Turkey for nine years as its own projects. The number of participants of Nature Education in 2006, which was the last year that these projects were carried out by TUBITAK, was 270 and the target group consisted of teachers, research assistants and the graduate students. The number of participants in 2006 was however too limited to create a wide-spread effect.

Publishing the other project call entitled 'the Science Camps and Schools' have common objectives with the project call for Nature Education. The general aim of the project call entitled 'Science Camps and Schools' is to introduce the relationship of science with daily life, show that experiments are not limited to laboratories; instead many experiments can be made with easily available materials from our homes and surroundings and to disseminate 'scientific thinking'.

TUBITAK had the experience of organizing a science camp. The number of participants of TUBITAK Science Camp was not sufficient to reach out the student from different regions of Turkey. Thus, TUBITAK had decided to evaluate the application of a project to improve the quality of nature education, to trigger the creative ideas on science camps and schools, to increase the number of participants and expand the target groups in a competitive way. In this sense, TUBITAK allocates money, $26 million in 2007, to motivate Universities and Governmental Institutions for designing and applying similar projects and started a programme for supporting science and society projects. The science and society support mechanism, established in 2007, consists of three stages. These stages are the application procedure, evaluation of projects and the project conclusion.

The application process begun with filling in the application form, examining the principles of finance and management and the regulations of the Science and Society Department. Then all the regulations are set by experts at TUBITAK. In the application form, the project is explained as a set of activities planned in a detailed way and the aim, scope, space, time and the target group play vital role in the evaluation by TUBITAK. The application form has three dimensions to be fulfilled by the project coordinator. The application form points to the fact that projects are evaluated in accordance with these three dimensions the method, how the activities can be done, the widespread effect, how many people can benefit from this project, the feasibility, the relative evaluation of the potential outcome and support demand. In addition, the applicant had to pay attention to the principles of finance and management. It determines the restrictions of who can (not) be the project coordinator, which expenses are met by TUBITAK, how one can form the budget, etc.

The evaluation of projects has also three steps; the projects are evaluated by independent advisors, experts from TUBITAK and the Commission of Science and Society Programmes. The advisors are chosen from the academicians who participated in TUBITAK's Science and Society activities voluntarily for ten years. Because it shows that these people have the vision and capabilities to popularize science and can be successful to prepare a new agenda for science communication. The experts from TUBITAK adopt multi-disciplinary approach that evaluate the project on the basis of objective criteria and control the truthfulness of the projects' within a scientific context. The project calls for the science camps and schools and nature education take the heed of the needs of multi-disciplinary approach in reference to Frascati Manual which includes definitions of basic concepts, data collection guidelines, and classifications for compiling Research and Development statistics in OECD member countries. (http://www.nsf.gov/statistics/randdef/iorg.cfm#oecd)

In other words, the project calls for the science camps and schools and nature education determine the branches of science that can be focused on in candidate projects, in reference to the Frascati Manual.

The experts from TUBITAK organize and construct workshops, festivals, camps on the field of the science and society. These experts are not only responsible for project evaluation which is the control of the planning and implementation of project activities concerning the objectives to be achieved, but also review the whole process of the project. The last authority of evaluation is the Commission of Science and Society Programmes, which consist of the Vice President of TUBITAK, head quarter of the department, academicians who have become experts on science communication and the scientific programme assistants.

The project applicants have the opportunity to make presentations prepared to reflect the value of their project in terms of science communication, the widespread effect and the feasibility, to the Commission of Science and Society Programmes This Commission plans to have meetings with the experts and evaluates the view of advisors. As the consequence of the intense meetings, the Commission takes the decision on the acceptance of Projects.

In 2007 the total number of the projects funded is 27 out of 83 candidate projects presented by Universities and Governmental Institutions. Total number of Science Camps and Schools are 14 out of 41. The target group of Science Schools and Camps are primary and secondary school students. The target group orientation are recognised by equity between men and women and low socio-economic status children are privileged in determining the selection criteria. Thirteen Nature Education projects were accepted out of 42 applications. The target group was varied from research assistants, teachers and graduate students to primary school students.

In 2007, the Projects of Nature Education were carried out in 11 different regions of Turkey.

As a consequence of Nature Education Programme, it is expected that the participants will develop a better understanding of local, regional, national as well as international environmental problems; will be able to discuss and provide alternative solution to global ecological crises; and will take action in their lives towards creating a more sustainable environment for future. The social and ecological relations within species are interpreted by participants in a scientific way of thinking.

Supported projects have reached up to 3500 people of different age groups and from various locations in Turkey. The support system provides opportunity for serving the majority of Turkey's population. In 2008, Science and Society Division also continued to increase and vary its projects to reach up to millions of people.

Conclusion

TUBITAK is working for dissemination of scientific culture in Turkey, both by organizing activities itself and funding other projects. The main purpose of designing its own activities is to maintain and improve the know-how for science and society activities and to be able to support high quality projects. The future strategy for forthcoming support programmes is to develop exhibitions for science centres, to raise science communication with computer and web based applications and to train the science communicators in 2008. In coming years new periodical calls will be published to accelerate dissemination of science literacy and the scientific culture among the public.

Reference

http://www.nsf.gov/statistics/randdef/iorg.cfm#oecd (accessed 14 Feb.2008).

Chapter 18

International Volunteer Programme

Nakano Hideo

Japan International Cooperation Agency (JICA),
Wesley College of Education, Kumasi, Ghana
E-mail: nh1886@yahoo.co.jp, nh1886@hotmail.com

ABSTRACT

Japanese science education reinforcement trends such as science festivals, super science school program, upgrading teaching and experiment skill for teachers are programmed to maintain Japan as a science and technology nation. All of these programmes are aimed at encouraging "Learning through experience". Children and teachers will be familiarised with new science materials and experiments through these programmes. These activities give one a feeling of accomplishment and make continuous virtuous circle of interest in science and technology. In African developing countries, accessible and affordable science and maths materials are few and far between. But we can apply low-cost and locally available products to make attractive science and maths materials. Introducing materials made from daily items can inspire local teachers and children to improvise science and maths materials and keep up with self sustainable development. Government stakeholders must improve educational material distribution system and teacher training system simultaneously.

Keywords: *Japan International Cooperation Agency, Japan Overseas Cooperation Volunteer, Rubber band, Straw, Science toys, Improvisation, International volunteer program, African developing countries.*

1. Current Science Education Promotion in Japan

1.1. Recovery of the Damage Caused by the Reduction of Contents of Textbooks

From 1980's Japanese schools have been suffering from school violence, bullying, school phobic children and cram education. In order to solve these problems the Ministry of Education, Culture, Sports, Science and Technology (MEXT) notified new government teaching guidelines for elementary and secondary education in December 1998. It was fully implemented from April 2002. According to the current guidelines 30 per cent of the basic school textbook contents were omitted compared to the previous guidelines. Flexible and liberal education was also induced.

In January and February 2002, MEXT carried out sample examination which targeted 450,000 pupils from grade 5 to grade 9. The average score of arithmetic, mathematics and social studies was lower than the previous exams which were carried out in 1996. The poor results and increased number of science phobia tendency among children made the MEXT change its tactics. In 2003 MEXT ended flexible and liberal school policy and shifted to revise and reinforce basic curriculum.

The reduction of science curriculum contents affected school children. While I was working at a private junior and senior high school in Japan from 2006 to 2007, I found serious defects. For example, some pupils could not use matches, have not seen siphon, did not know voltage of batteries nor domestic electric cable, have not burned off miniature torch bulb with using excess batteries. A number of pupils are not good at making science toys with scissors and pliers. School children were very familiar with these skills and knowledge few decades ago.

In addition, product liability law was put in place in 1995. The manufacturers forced customers to ban from disassembling their home electric appliances and toys. The industrial products became black box more than ever. In order to protect customers, manufacturers have made their products stronger and non-breakable. These days, children cannot learn structure and mechanism through disassembling toys. Japanese children used to learn how to handle fragile equipment and electric gadgets through checking their inner structure.

1.2. Samples of Activities to Enhance Science Education

In confrontation with science education crisis, a lot of science teachers, scientists and engineers in private sector and government organisation took action. They have started different programs to enhance science and technology education.

Science and Technology Basic Law

In 1995 Science and technology basic law was established aimed at building creative science and technology nation. In order to raise the research and development investment up to 1 per cent of GDP to the standard level of western nations, Japanese government decided to invest 24 trillion yen from 2001 to 2005.

Youngers' Science Festival

One of the most popular science promotion events is "Youngers' Science Festival" which started in 1992 organised by Japan Science Foundation (JSF). During the summer school holidays a number of teachers, scientists display and demonstrate their science materials and technological equipment in all scientific fields. Targets of this festival are children, parents and teachers from elementary school to high school. The festivals are held in every 46 prefectures and collect thousands of children to the each site. Most of the science materials which are introduced during this festival are made from daily products. They are affordable and easy to access in Japan even special high-tech materials such as rare-earth alloy magnets, shape memory alloy and replica grating sheets. How to assemble and how to use these science materials are printed in a booklet and distributed to every government and private schools without charge. Teachers can apply the new materials to daily classroom lessons and children can make the favourite ones for their projects.

241 science museums were registered to Japan council of science museums in December 2006. They are public and private institutions such as science museums, museums of national history, a mining museum, an agricultural museum, aquariums, volcanic science museums, space museums, energy museums and so on. These science museums are also supporting science education through having workshops, organising events for children and lots of retired engineers, scientists and school teachers are assisting their activities as volunteers.

Science festivals and museum workshops give more opportunities to children to learn through experience, provide chances to master how to use tools, how to measure, and the events also stimulate younger generation to be interested in science and technology.

Super Science High School

From 2002 MEXT started Super Science High School (SSH) project in order to upgrade science education among high schools. MEXT designates SSHs through judging proposals from aspirant schools. In 2007 a total number of authorised SSHs were 101 (80 prefectural schools, 6 state-run schools and 15 private schools). The annual budget for SSH project is approximately 1.4 billion yen. Main activities of SSH project are as follows

- Promote learning through experience and guided studies
- Develop and implement enriched curricula in science, technology and maths.
- Classroom teaching and presentations in English to prepare students for international activities
- Develop teaching methods and materials to enhance creative and independent thinking
- Participate in international contests
- Collaboration between high schools and universities to promote science education

Workshops for Science Teachers Promoted by MEXT

In order to update the skills and knowledge, Japanese teachers are recommended to attend workshops. Most of the workshops are held in summer holidays. I attended genetic recombination experiment workshop for high school teachers held at Shizuoka University in August 2006. This workshop was funded by MEXT as one of the activities of "Science Partnership Project". Through such workshops MEXT aims to propel adoption of new experiments to high schools.

Figure 18.1: Open School Day at a SSH

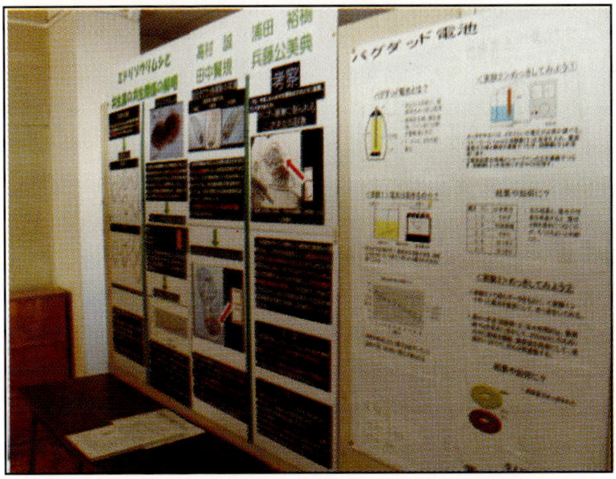

Figure 18.2: Student's Panel Presentations

The outline of the workshop was as follows (Figures 18.3–18.8).

2. Some Suggestions to Strengthen Science Education in Africa

2.1. Observation on Science Education in Zambia and Ghana

I came to Ghana in March 2007. I was looking forward to returning to Ghana after I had finished my first mission in Zambia. I worked at National Science Centre (NSC) in Zambia from April 2004 to April 2006 as a technical adviser. In March 2006, I attended a Science video project workshop which was organised by USAID in the capital city of Zambia. The purpose of the workshop was to select model lessons which we could record in video CDs and distributed to resource centres and schools. One of the teachers who prepared a lesson on electromagnet had the accurate aspect

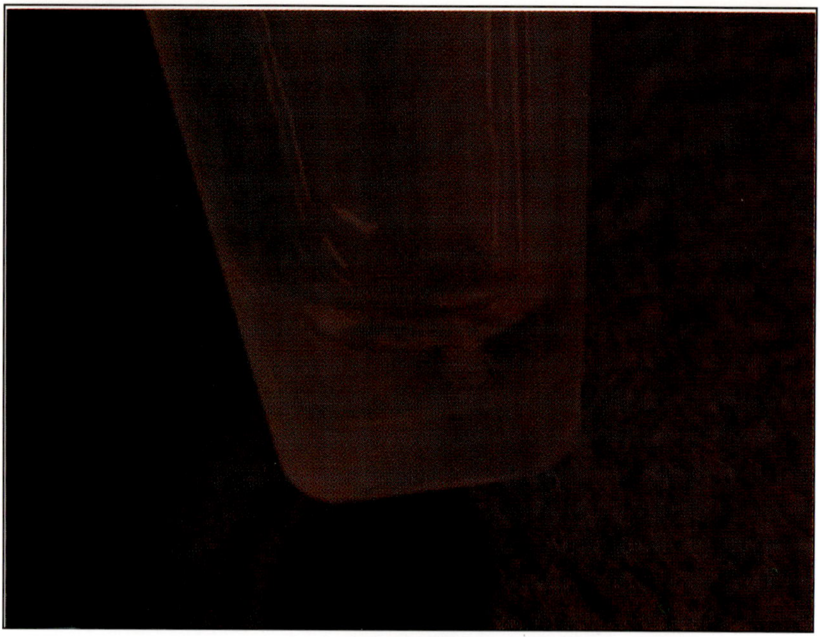

Figure 18.3

Experiment 1: PCR technique and DNA identification

Extracted DNA samples from shredded tissue like hair root or nail through de-protein process.

After enzyme treatments and several centrifugations, collected white threads like DNA precipitate.

Figure 18.4

Using Polymerase chain reaction proliferated target DNA sequence up to millions of time. The target were mitochondrial D loop and sequence which codes aldehyde dehydrogenase 2.

Figure 18.5

After three hour PCR reaction, implemented DNA electrophoresis using buffer and fluorescent dye.

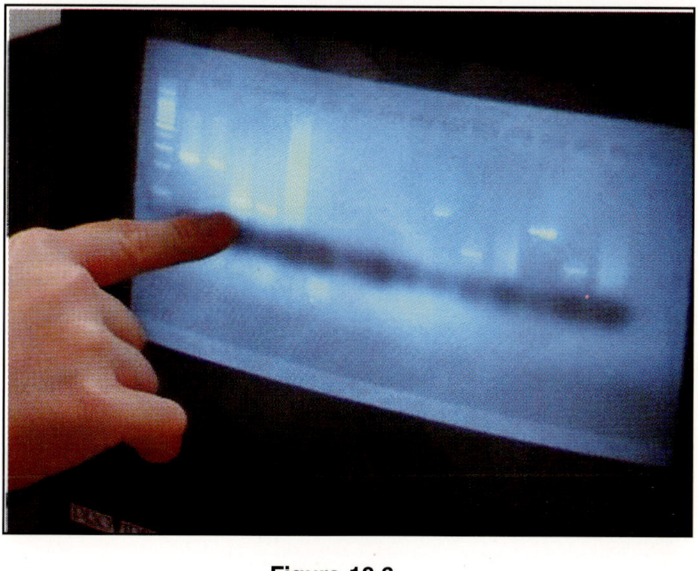

Figure 18.6

Under ultra violet lamp, identified target DNA by comparing with maker DNA

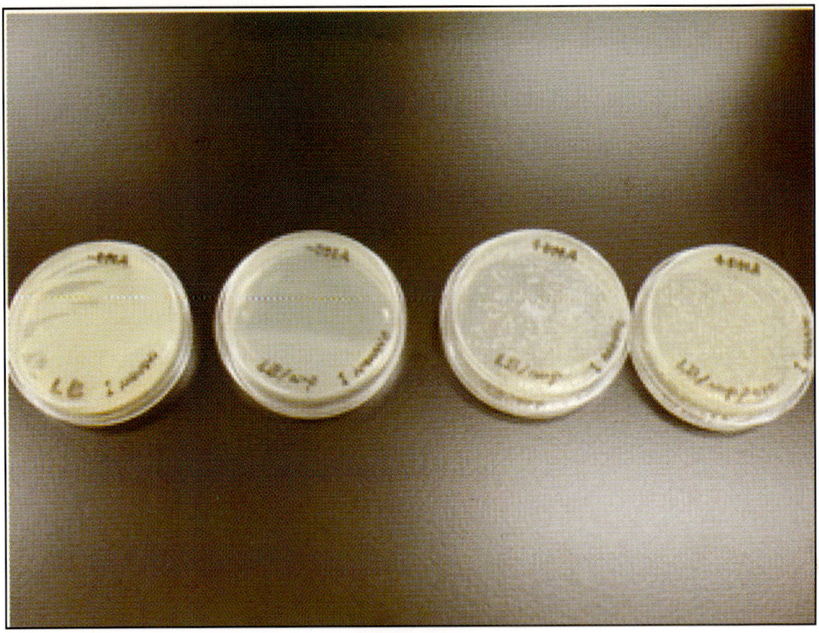

Figure 18.7

Experiment 2: Genetic transformation

Transferred pGLO circular DNA plasmid which has green fluorescent protein (GFP) gene into E-coli cells under existence of membrane destabiliser through heat shock.

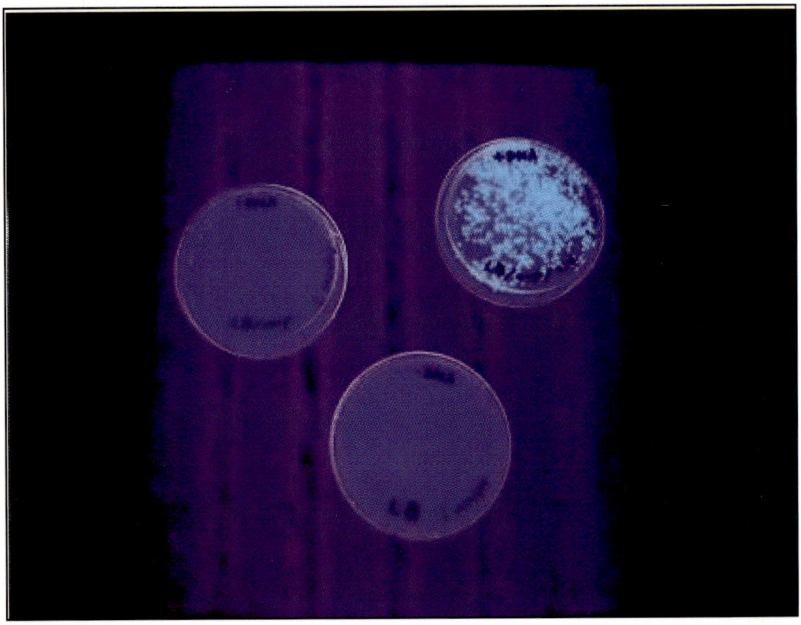

Figure 18.8

After cultivation, checked genetic transformation under ultra violet lamp

of theory but the practical did not work out properly. Because the textbook instructed to wind insulated copper wire up around an iron wood screw five times. After I listened to her explanation, I said, "Please wind up the wire as many times as possible, at least 50 to 100 times!" It was one of the typical episodes when we had workshops in Zambia. The teachers are well trained through books and they can explain in good English but most of them do not have enough practical exercises.

After I joined similar workshops in Ghana and lectured experiments to students, I came to understand that the basic background of practical experience among college tutors, junior high school teachers and students. Though I visited modern National Science Resource Centre in Accra last April and attended a workshop which was held there, I was easily able to estimate that at most schools in Ghana, science and maths educational environment are similar like in Zambia. I admire teachers' and students' language flexibility. They can memorise English textbooks thoroughly, but because of the lack of the practical experience it is very difficult for them to connect actual scientific phenomena to respective scientific explanations properly. Some difficulties in science educational environment were

- Teachers weight theoretical aspect of science too much without having experiments
- Lack of teachers' experience in the field of practical activities
- Lack of seeing and touching actual science materials makes it very difficult to imagine how to handle them even they are printed in the textbooks
- Tools and chemicals are not affordable
- Inadequate maintenance of facilities and poor access to essential consumables
- Rural schools are left too much behind
- Teachers tend to give up because of lack of materials.

2.2. Volunteers Activities in the Field of Science and Maths Education

From 1994 to 2005 JICA (Japan International Cooperation Agency) implemented 22 science and maths education projects in developing countries. For example in Zambia the JICA project is called SMASTE (Strengthening Mathematics and Science and Technology Education). Main activities of these projects are, in-service teacher training, publishing teacher textbooks, supporting school management, and revising curricula and textbooks.

From 1966 JICA dispatched science and maths teacher as junior volunteers (aged from 20 to 39) to developing countries. The total of dispatched teachers by the year 2006 was 2,146 to 57 countries.

1,364 volunteers were dispatched to Africa such as 366 to Ghana, 334 to Kenya, 208 to Zambia. They have a two year mission to work at local schools as grass-root supporters. In Zambia and Ghana junior volunteers are dispatched from elementary schools to colleges of education, and they also organise various activities like

Figure 18.9: Educational Material Workshop in Accra

Figure 18.10: Holiday School in Tamale

in-service teachers' workshops, educational material developing workshops, holiday school and experiment tour.

2.3. Suggestions

In order to conquer the above difficulties, NGOs have been assisting for years. While the recipients can receive financial and physical support, the projects look going very well.

Unfortunately, after the donors leave, lots of projects achievements are burnt into the ground. Such lessons imply that in African developing countries self sustainable development is the best tactics.

For the self sustainable science and maths educational development I propose some modification in the teacher training workshop system, educational material distribution system, and utilisation of low-cost and locally available educational materials like science toys.

Modification for in Service Teachers and Candidates Training

- Allow the teachers to make mistakes and give the participants enough days to complete projects after repeated trial and error.
- Find and apply local science handcrafts to the training curriculum.
- Check possibilities of introducing and developing of science cartoons.
- Distribute information for science and maths education through mass media and the Internet to encourage teachers constantly
- Enlighten and train teachers with mobile science centre.

How to Promote the Distribution of Essential Tools, Science and Maths Materials

Science centres or Ministry in charge of Science and Maths Education (MOE) should authorise (standardize) and inspect science materials which will be used in schools.

- Make safety regulation, specification standard and standard price of each science and maths material along syllabi and curriculum.
- At first the regulation should not be so strict. It shall be upgraded due to the growth of state's economy.
- Establish lists of educational materials and teaching aids which should be held at elementary and secondary schools, and resource centres.
- Dead line of achievement of the provision must be decided
- Science centres or MOE must propose financial support to fulfil the target.
- Promote and recommend production of standard educational materials to private companies.
- Science centre or MOE must inspect and examines the quality of distributed educational materials.

Samples of Low Cost and Locally Available Science and Maths Materials

As I demonstrated and lectured how to introduce low-cost and locally available science and maths materials in Zambia and Ghana at teachers' workshops, government schools and private schools, I found that they were very useful to encourage local teachers. These samples also gave hints to teachers how to utilise daily goods. Daily items like straws, plastic bottles, plastic bags, rubber bands, paper clips, bamboo sticks, and strings are common all over the world now. The equality of such products is uniform, reliable and the prices are affordable. Toys we can make in Africa can be applied to all the schools in the world.

I am going to introduce some samples below; I believe they are simple and exciting. Some people might have difficulties in making finished models because textbooks and guidebooks do not share enough space to explain how to avoid mistakes or how to adjust the final products properly.

Nevertheless, I would like to encourage teachers and children to keep on struggling till they complete their own projects. They can learn through try and error, can master how to handle assorted tools and these activities can connect theories with their true counterparts, scientific phenomena.

Plastic Straws are Versatile and Fundamental Ingredient

If you have a good pair of scissors you can make straw science and maths materials and toys comfortably.

Straws are very flexible and suitable to make various shapes (Figures 18.11–18.12)

Figure 18.11

Figure 18.12

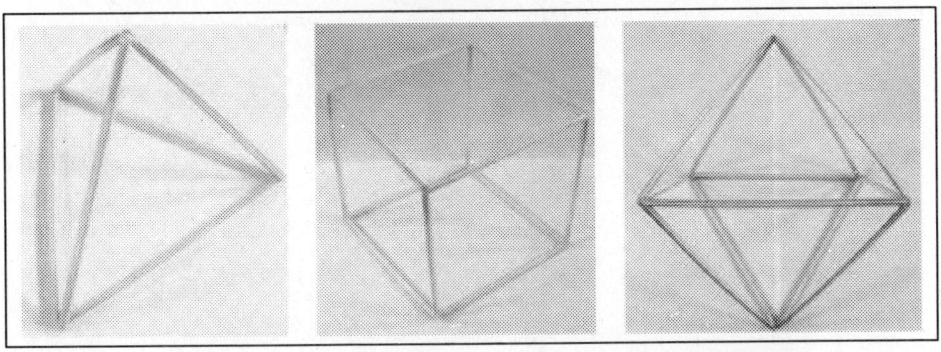

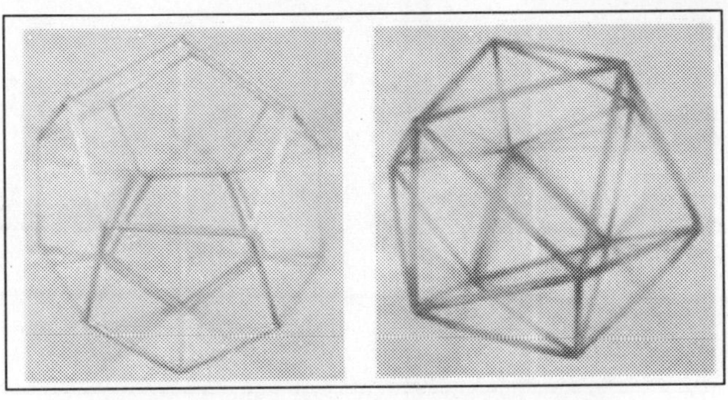

Figure 18.13: Straw Regular Polyhedrons

Because

- Straws are affordable
- Easy to collect in big numbers
- Several standardised straws are available in a country
- Easy to connect with making proper slits. Without glue or Cello tape slit straws are jointed with friction.
- Easy to replace and repair

Rubber Bands are Light and Powerful Engine for the Toys

Traditional thread spool runner is still very famous among aged people. Instead of using thread spool we can make plastic bottle runners. All we have to need is one or two plastic bottle(s) and straws or bamboo sticks, some rubber bands and a big bead or a perforated plastic disc. The bead or disc works as a bearing to reduce friction. When you wind up the handle in the figure, you can charge elastic energy in

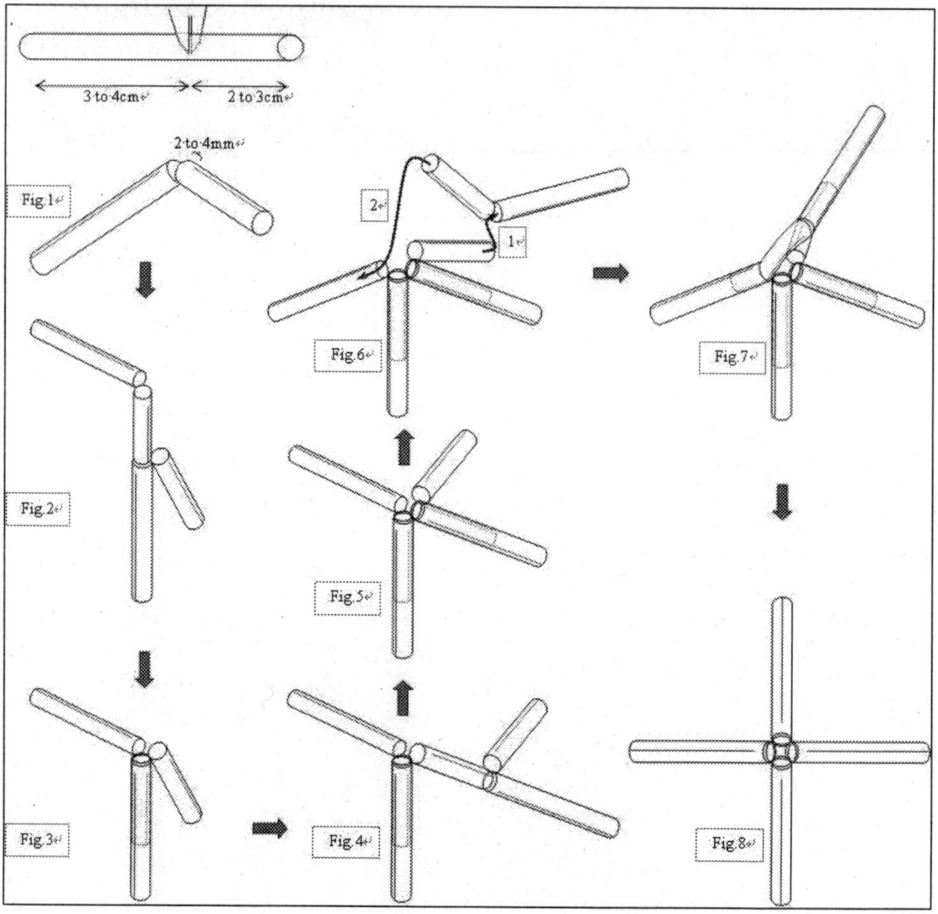

Figure 18.14: How to Make Four Arm Straw Joint

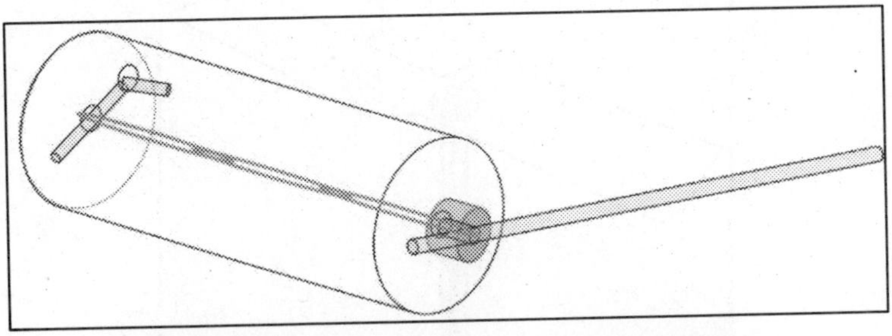

Figure 18.15: Spool Runner (Plastic Bottle Runner)

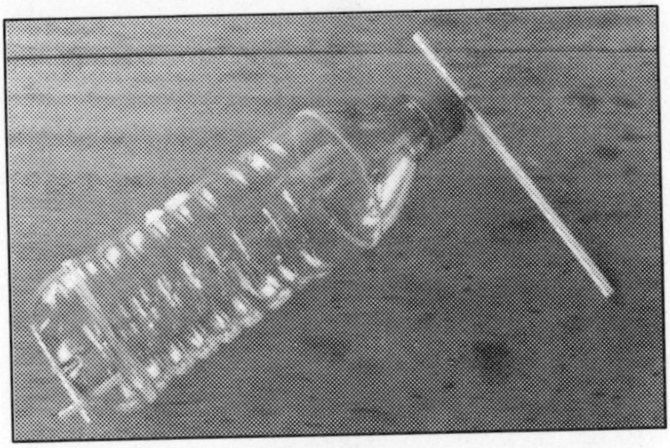

Figure 18.16

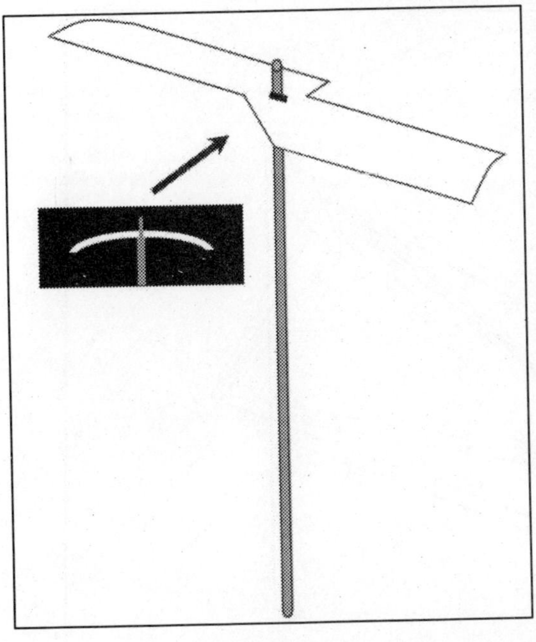

From a square prism shape bottle, you can cut out four propellers. From cylindrical bottle you can also cut out propellers.

With a good bamboo stick and the propeller you can make a very simple helicopter.

We can apply the plastic bottle runner and simple helicopter technique to make a propeller unit. Rubber bands are installed inside of a 6mm diameter straw. Hook and axis of the propeller is made from an office clip. A small bead also has an important role as a bearing. In addition, in order to make the unit you must master how to use long nose pliers.

As an application of straw polyhedron and the propeller unit, we can make a wind power vehicle.

Figure 18.17: Plastic Bottle Rotor Helicopter

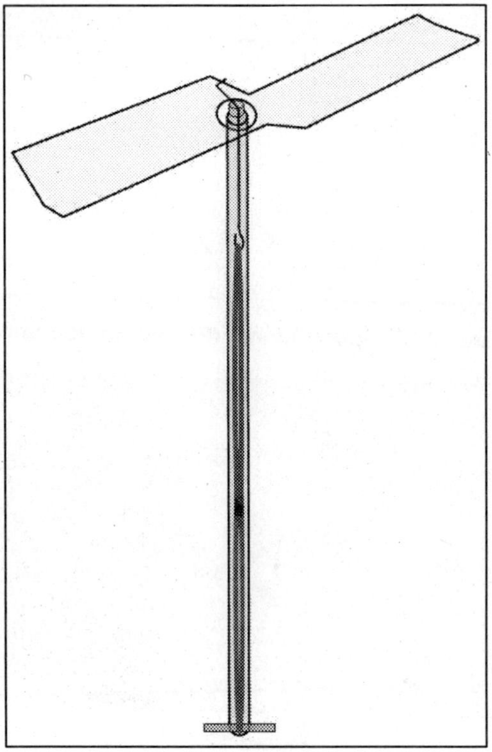

Figure 18.18: Propeller Unit

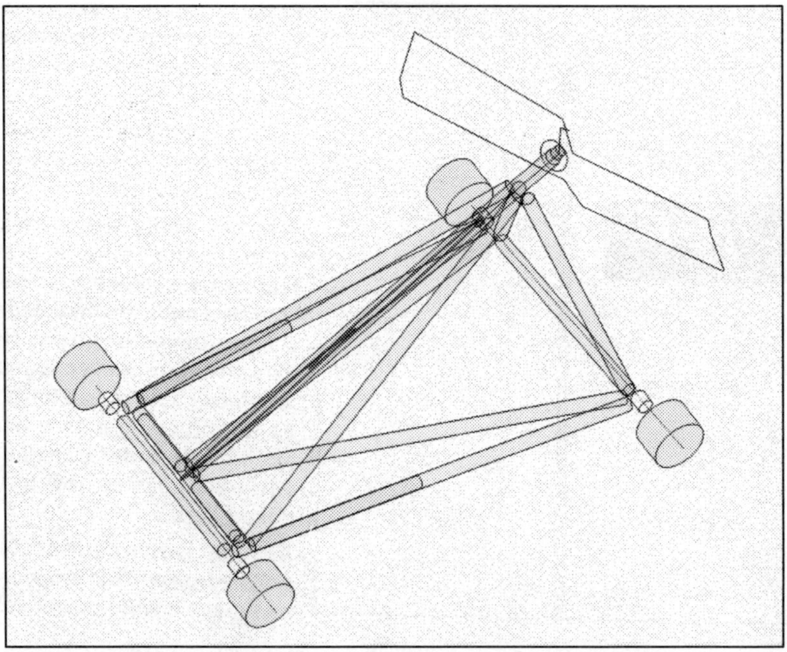

Figure 18.19: Rubber Band Propelled Wind Vehicle

extended and twisted rubber bands. Place the runner on the floor and release it. It runs smoothly.

One of the most exciting science toys is a rubber band helicopter. Using two light plastic cups, you can make upper and lower rotors. Upper and lower rotors have

Figure 18.20: Rubber Band Helicopter

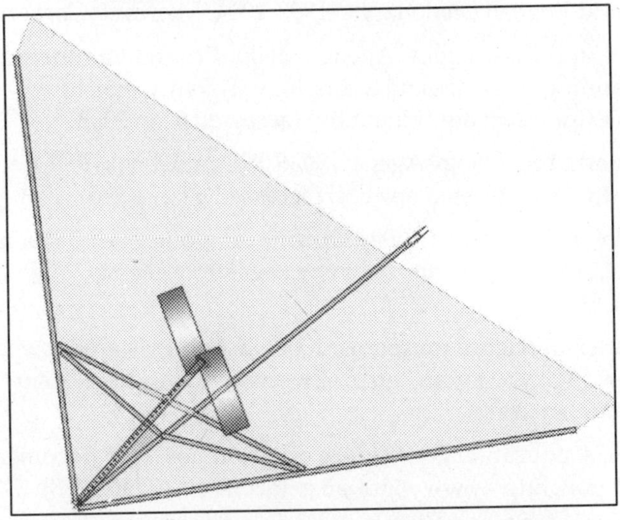

Figure 18.21: Rubber Band Propelled Straw Kite

Rubber band propelled straw kite is an application of straw maths models and propeller unit technique. You should make this kite as strong as possible and as light as possible. With only two small rubber bands it can fly up into the air. If it can ride on up-current of air.Please do not play with the kite near the road, railroad and electric cable!

These rubber band powered toys are good samples to demonstrate energy conversion and action and reaction.

different wing angles. The upper rotor rotates anti-clockwise and lower rotor, which is fixed to body straw, rotates clockwise.

Therefore this helicopter can fly up vertically into the air.

3. Conclusions

Japanese science education reinforcement trend such as science festivals, super science school program, upgrading teaching and experiment skill for teachers are programmed to maintain Japan as a science and technology nation. All of these programmes are targeting to encourage "Learning through experience". Children and teachers will be familiarised with new science materials and experiment through making mistakes. These activities give accomplishment feeling and it makes continuous virtuous circle of their interest in science and technology. This remark is not limited to my mother country. In industrial countries we can access and purchase huge range of science materials and tools. Unfortunately in African developing countries accessible and affordable materials are restricted. But we can apply low-cost and locally available products to make attractive science and maths materials. Introducing materials made from daily items can inspire local teachers and children to improvise science and maths materials and keep up with self sustainable development. Government stakeholders must improve educational material distribution system and teacher training system simultaneously.

References

Encarta® World English Dictionary © 1998-2004. Microsoft Corporation.

Japan Science and Technology Agency, 2004. Programme for promoting public understanding of science and technology, Tokyo, Japan. http://www.jst.go.jp/rikai/eng/for_museum/main.html (accessed 6 Jan 2008)

Japan Science and Technology Agency, 2006. JST guidebook (PDF document), Tokyo, Japan. http://www.jst.go.jp/EN/(accessed 6 Jan 2008)

Japan Science Foundation, 2004. Outline of youngsters' science festival (in Japanese),Tokyo, Japan. http://www.kagakunosaiten.jp/outline.html (accessed 6 Jan. 2008)

Japanese council of science museums, 2006. List of registered science museums (in Japanese), Tokyo, Japan. http://www.kahaku.go.jp/jcsm/list_print.html (accessed 6 Jan. 2008)

JICA, 2007. JICA educational assistance programmes (PDF document, in Japanese), Tokyo, Japan. http://www.jica.go.jp (accessed 17 Jan 2008)

Ministry of Education, Culture, Sports, Science and Technology, 2007. Announcement about 2007 Super Science High School Programme (in Japanese), Tokyo, Japan.

http://www.mext.go.jp/b_menu/houdou/19/04/07042001.htm (accessed 6 Jan 2008)

Abbreviations

- JICA: Japan International Cooperation Agency
- JOCV: Japan Overseas Cooperation Volunteer
- JST: Japan Science and Technology Agency
- MEXT: Ministry of Education, Culture, Sports, Science and Technology
- SSH: Super Science High School

Chapter 19

Traditional Technology Revived in Southern Most Malawi

Mike Gondwe[1] and Menno Welling[2]

[1]*Museums of Malawi, Box 30360, Blantyre 3- Malawi*

E-mail: mikemiggondwe@yahoo.com

[2]*Catholic University of Malawi, P.O. Box 5452, Malawi*

E-mail: wellingm@cunima.net

ABSTRACT

Tisunge! Lower Shire Heritage Centre with the aid of the Museums of Malawi has reintroduced the ancient art of spinning and weaving in southern most Malawi. This traditional technology was all but lost. Now some 20 families around Lengwe National Park can supplement their income with sales of hand woven cloth to national and international tourists. Moreover, the revival of this ancient craft has started a local reevaluation of ancestral knowledge and the often disparaged past.

Keywords: *Malawi, Museums, Cotton spinning, Cotton weaving, Culture and development, Traditional technology.*

Introduction

The Museums of Malawi is a government department under the Ministry of Tourism, Wildlife and Culture. Its mission is to collect, research, preserve and disseminate information to the public regarding natural and cultural heritage of Malawi. The Education Department of the Museums of Malawi educates people on the same through permanent exhibits and public programmes at Chichiri Museum in Blantyre and others across the country. The Museums of Malawi assisted *Tisunge!* Lower Shire Heritage Centre, a science centre in southern most Malawi, with the revival of the ancient technology of cotton spinning and weaving.

Background

Development is a priority of the Government of Malawi and poverty eradication is a Millennium Development Goal shared by the United Nations, the Commonwealth, NGOs and the people alike. Malawi has always been ranked among the poorest countries on the UN Human Development Index. In terms of GDP (Gross Domestic Product) per capita Malawi is indeed the poorest country of the world (UNDP, 2007/2008). The Museums of Malawi, in collaboration with other stakeholders, has embarked on a campaign to fight poverty through the lens of culture. Some of the issues it has concentrated on include food security, HIV/AIDS, malaria prevention and poverty. Slowly, there is increasing awareness that there is a direct link between culture and development.

Today the Museums of Malawi have become agents of change in development; they mirror events in society and become instruments of change by calling attention to actions and events that will encourage development in society. One of the

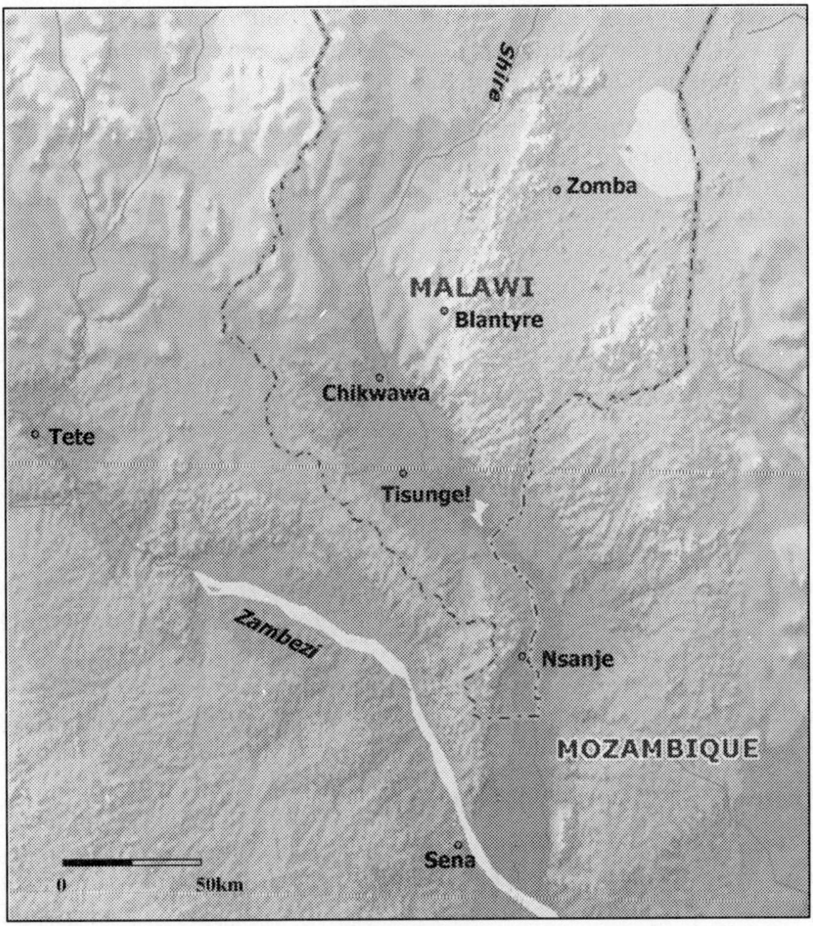

Figure 19.1

fundamental objectives of the Museums of Malawi is to educate and it is only the museum that has the capacity and ability to impart cultural education effectively as it houses the tools and materials for doing so in its collections.

What follows is an account of a project initiated by Mlambe Foundation, the Netherlands, which entailed the re-introduction of indigenous knowledge harboured in the Museums of Malawi to its original owners: the people of the Lower Shire Valley. It is the reintroduction of traditional spinning and weaving skills.

Cotton in the Lower Shire Valley

The Lower Shire Valley, the southern tip of Malawi, is known for its heat, mosquitoes, droughts and floods. Yet, the Lower Shire Valley has always been a hotbed of cultural activity.

From oral traditions and archaeology, we know of the original short-stature foragers (Robinson 1973, Welling 2000). From about the third century AD farmers were attracted by the fertility of the Shire flood plains (Robinson, 1973). These farmers brought with them knowledge of iron smelting and pottery making. Throughout the valley one can find the material proof of these early occupants in the form of small stone tools, potsherds and iron implements.

The Lower Shire had flourishing cotton weaving and iron smelting industries (Mandala, 1990). Its produce was traded to remote areas including that of the Mutapa Kingdom in what is now Zimbabwe (Monclaro, 1572). The cotton cloth and iron hoes were widely known for their fine quality. Centuries later, when Dr David Livingstone first set foot in the Lower Shire he made the same observation (Livingstone and Livingstone, 1865). Notwithstanding considerable Yao and Portuguese slave raiding, it was the same Livingstone and the process of colonization he put in motion that brought an end to this period of high cultural technological achievement. Machine-made colonial imports were much cheaper than the local products.

It is sad to observe the knowledge of these great traditions rapidly fading from memory; locally, nationally and internationally. In line with the current Malawi Government's major project to open up the Shire Valley, the Museums of Malawi, the Department of Antiquities and Mlambe Foundation attempt to revive and preserve the culture of the area for tourism. For this purpose, the Lower Shire Heritage Trust has been formed uniting all the stakeholders. This trust is constructing a Lower Shire Heritage Centre by the name of *Tisunge!*-let us preserve-at the entrance of Lengwe National Park behind the ancestral grove of Paramount Chief Lundu, who is the descendent of a powerful 17th century king (Schoffeleers, 1992). The centre will in the end include a small museum, a library, a children's club and outreach programmes to the community.

Traditional Technology Revived

Chichiri Museum in Blantyre has on display the Lower Shire traditional cotton spinning and weaving and occasionally its staff demonstrates the craft to visitors in the open air. The Museums staff was taught the skill by an elderly man from Ngabu, Chikwawa some forty years ago. In the Lower Shire itself the skills had effectively faded from practice and even memory.

Figure 19.2: Spindle Whorl and Carding Bow

Figure 19.3: Mchenga Women Spinning

The spinning process consists of the following steps. First, the separation of the cotton from the seeds, the cleaning of cotton lint using a small bow by plucking the string continuously through the cotton fluffs to remove dust and align the fibres. It is then turned into a roving by hand–twisting and is spun into yarn by a wooden drop spindle. If need be stronger double ply yarn can be produced by twisting the cotton yarn obtained after the spinning. All of this is a very tedious and time consuming process. Understandably with the availability of industrial yarn, the local product was quick to be replaced. An added advantage of the industrial yarn was the wide range of colours available. Most of the natural dyes used in the past just gave various shades brown. Already in the sixteenth century, but probably even long before, colourful Indian cloth was imported. The cloth was unraveled and the yarn reused in local production (Monclaro, 1572).

The weaving is done on a single heddle ground loom (Davison and Harries, 1991). That is to say yarn with alternative colours is stretched between two horizontal sticks that are attached to four wooden pecks some four meters apart. These lines form the warps. The weaver moves the shuttle with additional yarn back and forth through the warps, thus producing the weft. The weaver, sitting along side the warp thus starts at one end and slowly moving to the other as cloth is being produced. In this way, any length of cloth could be made. If two people work together, passing the shuttle back and forth, the width can be two arm lengths. The designs consisted mostly of stripes but also checkered designs were made, an example of which can be found in the repository of the British Museum, London which holds a total of 4 pieces

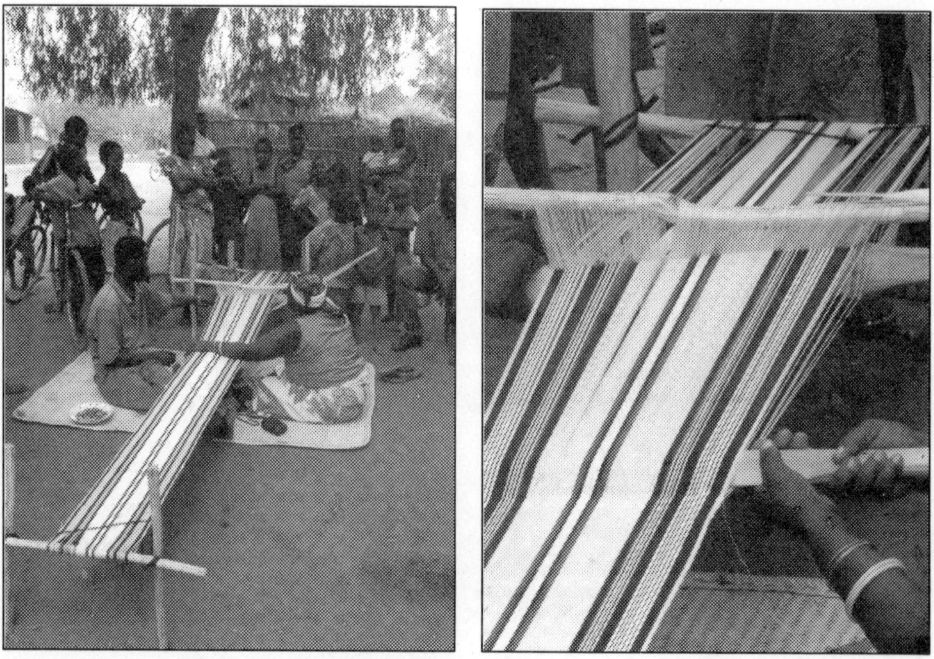

Figure 19.4: Weaving

of cloth and one loom collect by Sir Harry Johnston, the first British Consul to what is now Malawi. The pieces in the British museum are mostly large pieces (*chitenje*) meant to be wrapped around the waist. In order to economize on the expensive local yarn they are so loosely woven that they are see-through.

Through Tisunge! Lower Shire Heritage Centre, in 2006 Museums of Malawi was given the opportunity to re-introduce this indigenous knowledge for the first time since the 1930s.

Initially, the skill of weaving was taught to 10 groups from 5 different areas in Chikwawa district. Admittedly many of the participants took the lessons out of mere curiosity. Now, in a district where people by average earn less than a dollar a day, the hard working weavers that continued earn as much as $60 a month, as the hand-made, naturally dyed products are sold to tourist and others that come to visit Lengwe National Park. The experts even teach family and friends and improve on the colours. They continue to experiment with new natural dyes, making shades of red/brown, yellow and blue/gray, using local barks, leaves and flowers by boiling them with the yarn and adding salt to prevent running.

But what is more, as the awareness of traditional weaving spreads children learn to appreciate the skills and knowledge of their forefathers. This is a positive turn of events as since colonial times the local communities have been let to believe their heritage has little to offer.

Figure 19.5: Kanzibi CBO with their Weaving Certificates

Figure 19.6: Tisunge Shop

Figure 19.7: Dying

Tourism further helps this development. It triggers local reflection on the importance of this technology as people coming from far are taking an interest and are willing to pay good money for the cloth. Consequently, local cotton is given an added value which helps alleviate poverty in the area, while imparting young and old with appreciation of ancestral knowledge.

Thus the technique of weaving that was thought to have died out has been revived with positive effects to young and old. In effect, as it turned out, the technique had not died out completely. In 2007, one old man was found still weaving with his sons close to Lulwe mission, a remote area on Mozambique-Malawi borderland in Nsanje district. There, the skill had survived in the face of modern technology for the production of waist bands, locally known as *mcheka*. The bands are used by traditional healers in a spirit possession ritual called *malombo*.

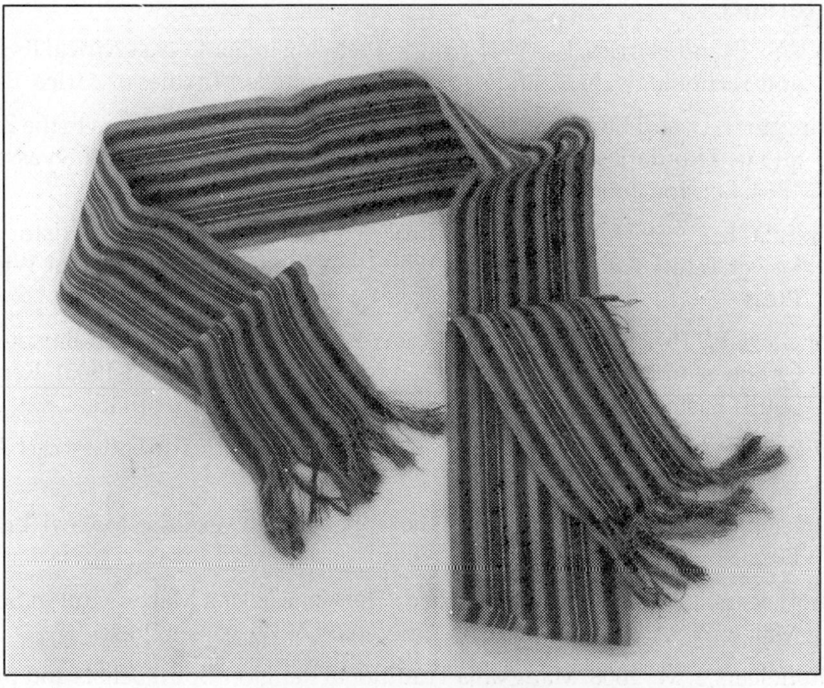

Figure 19.8: The Mcheka

Malombo is a ceremony for the appeasement of neglected spirits of departed relatives (Schoffeleers, 2008). Both healers and patients are donned with a *mcheka*, besides other paraphernalia. No substitute for the cloth can be used as only authentic hand woven cloth has the healing power. Apparently ancestral science continued to find a market in the context of ancestral religion. Muller and Snelleman (1892) collected a number of these *mcheka* waistbands during their Zambezi expedition in the 1880s. These bands are now in the possession of the Netherlands' National Museum of Ethnology in Leiden.

Conclusion

The *Tisunge*-Lower Shire Heritage Centre and the Museums of Malawi have revived traditional cotton weaving technology. They have helped the people at large, children and adults alike in appreciating their cultural heritage. Change is enhanced through informal learning and self-reliance instilled among the people. Community partnerships have been fostered and tourism is promoted.

Acknowledgement

On behalf of the people of the Lower Shire, we wish to thank ICCO/Impulsis, the Netherlands, for providing financial support for the construction of Tisunge! and its spinning and weaving programme.

References

Davison, P. and Harries, P., 1991. 'Cotton Weaving in South-east Africa: Its History and Technology'. In D. Idiens and K. Ponting (eds.) Textiles of Africa. Bath.

Livingstone, D. and Livingstone, C., 1865. Narrative of an Expedition to the Zambesi and its Tributaries:and of the discovery of the Lakes Shirwa and Nyassa, 1858-1864. London, John Murray.

Mandala, E. C., 1990. Work and Control in a Peasant Economy. A History of the Lower Tchiri Valley of Malawi 1859-1960. Madison, University of Wisconsin Press.

Monclaro, F., 1572. "Account of the Journey made with Francisco Barreto in the Conquest of Monomotapa 1572". In G. M. Theal (ed.) (1898-1903). Records of South-Eastern Africa. London, Clowes and sons. Vol. III: pp. 157-253.

Muller, H. P. N. and Snelleman, J. F.,1892. Industrie des cafres du sud-est de l'Afrique. Leiden, Brill.

Robinson, K. R., 1973. The Iron Age of the Upper and Lower Shire, Malawi. Lilongwe, Department of Antiquities.

Schoffeleers, J. M., 1992. River of Blood. The Genesis of a Martyr Cult in Southern Malawi, C. A.D. 1600. Madison, University of Wisconsin Press.

Schoffeleers, J. M., 2008. Mang'anja Traditional Religion. Spirit beliefs and practices in Malawi's Lower Shire Valley. Edited, annotated and introduced by M. Welling. Zomba, Kachere.

UNDP, 2007/2008. Human Development Report.

Welling, M., 2000. "Integrating Narratives of the Past: Challenge of Interdisciplinary Research and the Case of the Maravi States of East-Central Africa, c. 1500-1850 AD." Religion in Malawi 10: pp.29-35.

Chapter 20

Museums: Bases for Scientific Knowledge

Yvonne Ruwe Mulala

Lusaka National Museum, Lusaka, Zambia

E-mail: lusamus@zamnet.zm, myruwe@hotmail.com

ABSTRACT

Museums in Zambia are important sources of scientific knowledge and are promoting science in their communities, through their programmes, exhibits and displays. Looking at the large significance of Science Education and popularisation of Science and Technology in society at large, museums provide special scientific knowledge related to public programmes, such as "The Live Science Programme," at Lusaka National Museum, which brings out relevant, exciting and meaningful learning experiences to its audiences. The scientific knowledge creates awareness to the general public to appreciate science as part of their basic way of life. The National Museums Board (NMB) mission is "To Establish, Manage and Develop National Museums on a sustainable basis for the preservation and interpretation of movable heritage for the benefit of the present and future generations." There are four National Museums and two private ones in Zambia. In line with its mission, NMB assigned themes to its respective museum entities in 1996 through the restructuring process. These themes ranged from Cultural History to Natural History. Museums in Zambia use these different themes to promote scientific Knowledge to its visitors, through exhibits, displays and educational programmes and so far have become popular and shown to be important sources of scientific Knowledge. This paper looks at how Museums in Zambia are important sources of scientific knowledge and how they are promoting science through their exhibits and displays. It will also look at the creation of a Science and Technology Museum, as a way of raising awareness on research and stimulating the public for interest in science. Such a museum will attract more visitors by keeping in mind the personal, societal, economic, cultural and political values. The paper will further look at how scientific knowledge is being utilised by the Museum audience through specially designed educational programmes and activities.

Keywords: *Zambia, Museums, Exhibits, Educational programmes, Scientific knowledge, Science and Technology, Scientific temper.*

Introduction

Zambia is a Land Locked country and has a population of about 11 million, with a surface area of 752, 615 square Km. It has a wealth of natural resources for tourism, including the Victoria Falls, considered to be one of the greatest natural sites in the world and museums are one of the major tourism assets of Zambia.

The core functions of museums in Zambia are to exhibit, manage and develop staff, research, collect, document collections and conduct public programmes. The four National museums in Zambia are: the Livingstone Museum in the Southern part of Zambia; Moto Moto Museum in the Northern part of Zambia; Copperbelt Museum in the Central part of Zambia; and Lusaka National Museum in the Capital City of Zambia. The two private museums are the Nayuma Museum in Mongu, Western part of Zambia and Choma museum in the Southern part.

The four National Museums fall under the National Museums Board of Zambia (NMB), which was established in 1966 by an act of parliament CAP 174 of the laws of Zambia. The National Museums Board is a government grant-aided body in the Ministry of Tourism, Environment and Natural Resources. In addition, NMB has gone a step further in proposing the establishment of a museum of science and technology. All these museums have taken drastic measures in promoting scientific knowledge to the society and are playing a major role in being important sources of scientific knowledge.

History of Museums

The word museum has classical origins. In its Greek form, *mouseion*, it meant "seat of muses." The word museum was revived in 15th Century in Europe, to describe the collections. By the 17th Century, the word museum was being used in Europe to describe collections of curiosities. In 1675 the collections, having become the property of Elias Ashmole, was transferred to the University of Oxford and the building constructed to receive it, opened to the public in 1683. This became to be known as the Ashmolean Museum.

From this moment on, other buildings such as the British Museum were established to preserve and display a collection. Use of the word museum during the 19th Century and most of the 20th Century denoted a building housing cultural material to which the public had access. Open-air museums, comprising a series of buildings preserved as objects, and eco-museums, comprising of large outdoor environments, were also introduced as new ways to display objects, cultures and concepts.

From the different types of museums, there are science museums which have an important role in preserving the material evidence of scientific and technological endeavour. Other science and technology museums are more concerned with demonstrations rating science and its applications, and in these museums, the preservation of objects is emphasised.

Museums Promoting Change through Science in Zambia

Today most museums in the world are centres of conservation, study and are a reflection of heritage and culture. Museums are open to the public and their primary functions are to acquire, preserve, research, interpret and exhibit, for purposes of study, education and entertainment, material evidence of the people and their environment. In this case, museums portray the key elements of education, science and culture.

Museums in Zambia are promoting science through their exhibits, and specially designed science related public programmes. In the past few years, museums in Zambia have proved to be important institutions promoting scientific knowledge to its community and the results have tremendously shown that museums contribute to the development of education, science, technology and culture in Zambia. A few examples can be sited from the various museums.

The Livingstone museum was officially opened in 1934 and is the oldest and biggest of the four national museums. It has a Natural History section comprising the Mammology section, Ichthyology section, Entomology, Botany and Herpertology and has in total a collection of more than 20 000 specimens. Each of these sections is promoting scientific advancement to the community. Research is carried out in the respective sections, pertaining to the various disciplines and the preservation of specimens, their relationships to the environment and their economic value in terms of tourism is determined. In the Mammology section, Bi-annual bird counting is done in the Victoria Falls area as well as in the National Parks. Research is also carried out in the vegetation surveys of the park together with Zambia Wildlife Association (ZAWA) and the research findings are then communicated to the public through exhibitions and specially designed educational programmes. There is also a Wildlife Club which the mammologists are involved in with the schools and the community at the museum.

In the Entomology section, experiments on insects are performed, such as termites (isoptera) where the insects are kept under artificial conditions in the laboratory for study purposes and also to try and rear the termites in a museum laboratory. In addition, experiments on the reproductive and social behaviour of the different insects like the white-browned sparrow weaver are done. However, the limitation in advancing the department is that there's need to retain more skilled and trained man power, to fill the different sections. The other sources of scientific knowledge at Livingstone Museum are the natural history permanent exhibitions in the museum showing the diorama of food chain and a biodiversity of wildlife as well as the environment.

Following the National Museums Board of Zambia's restructuring plan of 1996, the Copperbelt Museum was assigned for Science and Technology development. The Board chose a thematic approach for each National Museum so as to widen the heritage specialisation in the public museums in Zambia and this plan has been included in the Fifth National Development Plan. Because of its advantageous location, the Copperbelt Museum was chosen to capture, preserve and express the development of Science and Technology in Zambia. The need for a Zambian museum of Science and Technology cannot be ignored. A consultative survey carried out revealed the desire by various stakeholders as well as other different groups and organisations, to create a museum of Science and Technology. During the conducted study, about thirty one institutions were interviewed and almost all representatives of organisations interviewed stressed the fact that the creation of a science and technology museum was needed forthwith. Zambia is a mining country and this museum will be a centre for reference on Science and Technology by the community and public at large. The Ministry responsible for heritage issues expressed the Government's commitment to the idea. This is an indication that the idea and desire has always been there but could not take off for various reasons such as lack of financial resources.

The museum of Science and Technology is expected to serve as

- A reference for Science and Technological innovations in the country
- An institution that promotes science and technology especially among the young
- Documentation centre of the various technologies applied in the various industries in the country
- Source of knowledge in the fields of science and technology.

The planned Museum of Science and Technology will incorporate both a Science Centre and a conventional museum. The Science Centre Departments objectives will include contributing to the improvement of the environment in which science, mathematics and technology are taught; promoting artisan skills; promoting awareness on science based careers; increasing public understanding of science; and uplifting minimum level of technological literacy. The museums objectives on the other hand will be to document innovations and developments in science and technology in Zambia; to raise awareness among the people of Zambia of the role of science and technology in national development; to capture other aspects of history of the people of Zambia that was directly influenced by technological innovations.

In other words, the Composition of the Museum of Science and Technology will include aspects on:

Science

- Which will include Earth sciences (Geography, Geology, Palaeontology, Archaeology, etc.) showing our changing world.
- Physical Sciences (Chemistry and Physics) where visitors will be allowed to discover and make some investigations.

- Life sciences (molecular and cell biology, and animal, plant and environmental sciences) which will show our life.
- Mathematical sciences

Technology

- Which will include manufacturing, processing, architecture, construction, telecommunication, mining, quarrying, and agriculture, only to mention a few.

The museum will also become an important cultural facility in Zambia's tourism arena and will attract tourists from all over the world. As a marketing strategy, it will be linked to the national and regional tourists trails, meaning that this will include networking with the copper mines on the Copperbelt, hotels in Ndola and airlines, as part of a package.

However, the challenge still remains to source for adequate funding for construction of this museum and also to meet the outlined objectives. The proposed sources of funding would be through the government grants (GRZ), international grants, private donations, local authorities and others. This museum would also need well qualified manpower, specialised in science and technology. This means that the museum will have to employ and train qualified manpower. In an effort to promote and popularise science and technology among the young, adults, and the public, the museum will have an outreach programme of the science centre section. This will include mobile services in terms of exhibitions, presentations, workshops, in order to reach out to the people in rural areas, as well as the visitors who cannot access the museum.

The Lusaka National Museum is also promoting change through science by offering scientific knowledge to the public, through its science related educational programmes and activities. Through its Education Department which strives to make the museum a forum for evoking and influencing change in the community, the museum runs a special programme called "They Lived Science." This programme was started in 2003 and focuses on science and technology in cultural artefacts. The project was sponsored by the European Union (EU) through the Cultural Sector Support Programme. It was intended to project the Past Africa's contribution to World Civilisations from Zambia's point of view, through, objects, specimen and value systems that portrayed scientific norms and protocols from indigenous practices and knowledge systems.

The objectives of this programme are as follows:

- Develop and manage an interactive (innovation and creativity) programme for the public on 'Environmental Sustainability' with regard to natural and cultural heritage (Indigenous Knowledge and Practices).
- To make an assessment of the demand and options for a young people's driven environmental awareness system in Zambia
- Promote children's awareness of the importance of both tangible and intangible heritage as thus can contribute to their scientific and technological learning and understanding.

- Contribute to the Environmental Education in Zambia and beyond.
- Conduct research, seminars and workshops, mount exhibitions/display and produce publications for academic and general public consumption.
- Develop a framework to analyse the scientific and technological principles embedded in tangible and intangible heritage.
- Institute an annual exhibition on environmentally friendly products based on indigenous knowledge systems and practices.

The thematic area identified to accommodate the project was 'applied technologies' as in relation to environmental sustainability. The project was therefore to be conducted through four main stages, which are Research, Workshop, Exhibition and Educational Outreach programme.

Results and Discussions

Records have shown that popularising and promoting science and technology is cardinal to the society and to everyday activities in all walks of life. Museums being custodians of heritage and culture of human kind play an important role in telling the story of the science and technology aspect of mankind. The science related programmes offered in the various museums in Zambia cater for all types of groups, ranging from the children to adults. To the public, the museum is considered to be a centre of fundamental knowledge of their own Zambian or African being. The public has so far shown interest and commitment in working with the museums to promote scientific knowledge. A visit to the museum reveals a typical display of exhibited collections of objects and its related information. These collections have knowledge embedded in them on science of (and) purpose as experienced by the people who made them.

Museums in Zambia should therefore not be an exception among important institutions promoting scientific knowledge. Museums being sources of scientific knowledge should work along side with the science centres as well as other science and technology institutions, in order to spearhead the advancement of science in the country. The results for instance, from the programme "They Lived Science" at Lusaka National Museum have been the publication of the researched thesis by the Education Officer entitled 'They Lived Science–Monkey Orange and Traditional Ecological Knowledge in Zambia.' Secondly the award of the exhibition certificate awarded to the collaborating partner, Miapen Zambia Crafts by National Science and Technology Council, for obtaining first prize industry in the Science and Technology exhibition in 2007. Thirdly, a Memorandum of Understanding was signed in January, 2008 between Lusaka National Museum and MIAPEN Zambia Crafts, on Heritage and Environmental Sustainability. Several workshops have been conducted involving different groups such as women, children and adults, and temporal exhibitions have been put up at the museum.

Conclusion

From the discussions above, the role of museums has clearly indicated that there is need to consider museums as important sources of scientific knowledge and contribute to national development as well as provoking change in the community. Museums, along with science centres, as well as other institutions offering scientific knowledge should be supported by both the government as well as the public at large. This can be done through sharing experiences and knowledge with experts from other countries all over the world and gain new ideas. This will help everyone actively involved in research, extension in science popularisation, communication and working in science centres, science museums and similar projects, to acquire the necessary skills and knowledge required for national development.

References

Mudenda, G.S., 2006. Status of Museums in Zambia: Are Museums Mirrors or Warehouses. In: The Enhanced Role of Science Centres and Museums in Development Strategies. New Delhi, India, pp. 116-124.

National Museums Board of Zambia. 2006. Feasibility Study Report for the Development and Construction of a Museum of Science and Technology. Ndola, Zambia.

Salasini, C. M., 2007. They Lived Science–Monkey Orange and Traditional Ecological Knowledge in Zambia Thesis, University of Tromso, Norway.

The New Encyclopaedia Britannica, 1768. Volume 24, Chicago, pp. 478.

The Livingstone Museum Newsletter, 1995. Volume 3, Number 1, Livingstone, Zambia.

Chapter 21

Role of Science Parks in Nigeria

M.A. Jolaoso[1], R.N. Ihenacho[2], D.B. Ayo[3] and P.A. Onwualu[4]

[1]*Investment Promotion and Consultancy Services Division, RMRDC, Nigeria*
E-mail: jolaosomoyo@yahoo.co.uk
[2]*Planning and Information Services Department, RMRDC, Nigeria*
E-mail: raynekky@yahoo.com
[3]*Industrial Plant and Equipment Department, RMRDC, Nigeria*
E-mail: danielayo@hotmail.com
[4]*RMRDC, Nigeria*
E-mail: ceo@rmrdc.org

ABSTRACT

The Raw Materials Research and Development Council (RMRDC), an agency under the Federal Ministry of Science and Technology (FMST), was established in 1987, with the mandate to promote, support and expedite industrial development and self-reliance through optimal utilization of local raw materials as industrial input. In pursuance of its mandate, RMRDC set up the Raw Materials Display Centres (RMDC) across the country (36 States). The Centres which plays vital role in local raw materials development has amongst its principal objectives to, create awareness of the existence of important raw materials locally; and stimulate interest in the use of these local raw materials as feed stock for both old and new industries. They offer at glance information to researchers, investors, entrepreneurs, policy makers, students and the general public with the aim of promoting investments in resource-based industries. The RMRDC in 2006, re-modified the RMDC and changed them to the Raw Materials Information Centres (RMIC). The RMIC as conceptualized is a multi-media resource (science) centre that combines both electronic and pictorial display of raw materials and provide data on all matters relating to raw materials sourcing, development and utilization. It also provides a critical linkage between the Research Institutes and end-users of R&D information, through data sharing via the Raw Materials Information System

(RMIS) developed by RMRDC. RMRDC hopes to key this programme into the broader one of Science Parks establishment, which is being coordinated by the Federal Ministry of Science and Technology (the parent Ministry).

Keywords: *Nigeria, Raw Materials Research and Development Council, Raw Materials Display Centres, Raw Materials Information Centres, Raw Materials Information System, Science Parks, Federal Ministry of Science and Technology.*

1. Introduction

The Raw Materials Research and Development Council (RMRDC) an agency under the Federal Ministry of Science and Technology (FMS&T), was established in 1987, with the national mandate to promote, support and expedite industrial development and self sufficiency through optimal utilization of local raw materials as industrial input. In pursuance of this objective, the Council set up Liaison (State Coordination) Offices and along with them, Raw Materials Display Centres (RMDC) in the 36 States of Federation of Nigeria and its headquarters is located at the nation's Capital, Abuja, FCT. The RMDC which plays vital role in local raw materials development has amongst its objectives the following:

- Stimulate interest in the use of these local raw materials as feedstock for both old and new industries;
- Focus attention on results of on-going and completed R&D on the utilization of local raw materials and enhances their early commercialization;
- Provide vital information to local and foreign investors interested in local sourcing of raw materials; and
- Promote R&D activities that will eventually lead to better quality and efficiency of productive capacities of local technologists.

The Centres as conceptualized in 1990 were designed to showcase the domestic raw materials resources, the intermediate and final products obtainable from them, investment opportunities, process flow charts, and prototype/model process plant of important raw materials endowment of the State that needs to be promoted at that point in time (RMRDC, 1991). It offers at a glance, information to researchers, investors, entrepreneurs, policy-makers, students and the public with the general aim of promoting investment in resource based industries. The displays are arranged around the Manufacturers Association of Nigeria (MAN), 10 sectoral themes; *i.e.*

- Food, Beverage and Tobacco
- Wood and Wood products
- Pulp, Paper and Paper products, Printing and Publishing
- Textile, wearing apparel, leather and leather products
- Domestic and Industrial plastics, Rubber and Foam
- Chemicals and Pharmaceuticals
- Base metals, Iron and Steel

- Non-metallic minerals
- Motor vehicles and Miscellaneous
- Electrical and Electronics

In 2006, RMRDC as parts of its internal re-organization, re-modified the RMDC and changed them to the Raw Materials Information Centres (RMIC). The RMIC as conceptualized is a multi-media resource (science) centre that combined both the physical (as in RMDC), electronic and pictorial display of raw materials and provides in addition, data on all matters relating to raw materials sourcing, development and utilization. It also provides a critical linkage between the Research Institutes (RIs) and the end-users of R&D information through data sharing via the Raw Materials Information System (RMIS) developed by RMRDC (Thompson, 2007). They therefore have in addition to the traditional RMDC, the RMIS and the electronic/pictorial display and the all important V-SAT facilities providing the linkage among all the RIs in the Country to the outside world.

2. Raw Materials Information System (RMIS)

RMRDC identified from the onset that lack of information on raw materials was a major hindrance to industrial development in Nigeria. It therefore became imperative for her to evolve a framework that would ensure proper inventory and documentation of the locally available raw materials and provide easy accessibility of such information to end-users.

To actualize this, a functional databank called the RMIS was conceived for the processing, storage and dissemination of information on raw materials, (quantity, quality, location, reserve estimate, etc), process technology, equipment selection, new product formulation/development, efficient enhancement of indigenous R&D activities, standardization/certification, testing and quality control, productivity empowerment, consultancy services from numerous and periodic techno-economic surveys conducted by the Council. The RMIS was developed by the Council in collaboration with the United Nations Development Programme (UNDP).

Overall objectives of RMIS amongst many others include the following:

- Serves as Centres for design and implementation of raw materials information system, collection, storage and utilization of information on such materials.
- Develop a data bank on raw materials sourcing, processing technologies, utilization and development.
- Exchange information with other national, regional and international organizations involved in the processing and utilization of industrial raw materials for the purpose of dissemination of information to end-users.
- Utilize the networking and data communications techniques to interlink with public and private centres of information and organizations involved in the processing and utilization of raw materials.

- Enables RMRDC to achieve its goal of being a national, regional and international focal point for dissemination of information on raw materials for both the public and the private sectors.

3. Very Small Aperture Terminal (V-SAT) Installation/Wide Area Network (WAN)

The implementation of the RMRDC V-SAT/WAN project commenced in 2002, with the first installation at the headquarters of RMRDC (Abuja). The success of the first installation led to the immediate expansion of the network to sixteen (16) remote locations, which accommodated some Research Institutes under the Federal Ministry of Science and Technology as well as RMRDC Zonal Offices. Subsequent expansion was carried out in 2005 with additional twenty-six (26) remote locations connected to the network.

Furthermore the project was expanded to cover other locations under the Public Service Network (PSnet). This project is to house all scientific and technological information in Nigeria with major focus on R&D. It is envisaged that the linkage will remove duplication and repetition of research activities by different organizations as well as make available information on research findings to relevant organizations. The projects have amongst many others the following objectives to:

- Serve as integral part of the Public Service Information Network (PSnet).
- Serve as an important tool for the Research and Development activities in Nigeria.
- Facilitate access to research information and materials and enhance dissemination and sharing of research findings

Table 21.1: Distribution of V-SAT Installation

Sl.No.	*Location*	*No. of V-SAT*
1.	RMRDC (Headquarters)	1
2.	RMRDC (State Offices)	21
3.	Research Institutes	16
4.	Minister's Office and others	5
	Total	43

RMRDC Website (www.rmrdc.org)

Abridged version of RMIS is also made available on the internet via the RMRDC website. The website was installed in 2002, with the sole purpose to promote activities of the Council on the information superhighway. The website provide information about Council's mandate, activities, key projects and programmes, raw materials inventories on Local Government Area basis (3rd tier of Government in Nigeria), briefs on resource based projects, Council's publications, list of equipment/machinery fabricators etc. At periodic intervals, the website which is data driven is updated to reflect information on new activities of the Council. In addition the RMRDC also coordinates the website for the G.15 countries project. The website provides avenue for interested international investors to visit and make enquiries in various areas of interest especially those bordering on raw materials development and utilization in Nigeria.

4. The Science Parks Establishment Programme of the Federal Ministry of Science and Technology in Nigeria

One of the major or cardinal programmes of the Federal Ministry of Science and Technology (FMS&T) in 2006 was the establishment of Science Parks (SPs) in Nigeria.

The programme is based on the fact that the Ministry recognized that S&T must take its proper position as the real engine of economic development. One of the strategies of FMS&T to attain this objective is to reposition S&T innovation system to propel the Nigerian economy back on track is the establishment of Science Parks in the Country.

The world over, Science Parks have been recognized to have positive impact/ bearings on national economies as they:

- Stimulate and effectively manage the flow of knowledge and technology amongst universities, R&D institutions, companies and markets.
- Facilitate the creation and growth of innovation based companies through incubator and spin off, process and
- Provides other value added services together with high quality space facilities.

Science Parks which are specialized industrial parks are managed by specialized professionals, whose main aim is to increase the wealth of its immediate community and its country at large by promoting the culture of innovation and competitiveness of its business and knowledge based institutions. SP's facilitate interactions between education, research and technology development (FMS&T, 2007).

Nigeria had made numerous attempts in the past, aimed at steering the State/ zonal/regional and national economies to be technologically driven with varied levels of success and failures, such attempts include: the Akwa-Ibom Park, Industrial Development Centres (21 States),Technology Incubation Centres(6), Free Trade Zones, Industrial Parks and the FMS&T Advanced R&D Sheda Science and Technology Complex (SHESTCO).

As conceptualized by FMS&T, the Science Parks will serve as a strategically dedicated sector, a research, business, technology and training centre all in one. The parks will also have supplementary uses such as Congress Centre, schools, sport-ground, museums and recreational centre. They will also provide advanced services such as R&D process centre, technology transfer and commercialization centres.

Deliverable expectations from the establishment of the centre in Nigeria are; efficient, linkage of Science and Technology, facilitation of measurable performance of academic institutions, R&D activities and response/contribution of the organized Private sector, dynamic evolution of desired technology, timely commercialization of innovation, fast tracking and development of incubators and SME's, and lastly add value to regional academic output/input.

Taking into consideration that mere copying into-to an infrastructure like Science parks which seem effective and functional in other countries may not totally serve our purpose, certain "realities on ground" must be featured into the planning,design,establishment and management of SPs at the earliest time otherwise the programme might fail as others, have.

The challenges are: large population of 140 million citizens, a burgeoning democracy since 1999, a multi-ethnicity of society with over 300 ethnic groups, long history of military dictatorship, over 70 per cent of population dwelling in rural areas and a brain–drain phenomenon since early 80's.

The strength includes the national vision target of being one of the top 20 leading economies of the world by 2020, healthy national Science, Engineering, Technology, and Innovation (SETI) system, with over 80 Universities, 60 Polytechnics, over 60 colleges of Education and over 120 Research Institutes and un-estimated innovation centres in the private sector, healthy budgets for development, with over 250 billion technology oriented components of yearly national budget, a wide quantum and quality of documents of documented and undocumented R&D results and private inventions/innovations covering low, medium and high scale technology and a respectable foreign-reserves of about $50 billion, post Debt relief from the G. 8 countries.

The FMS&T in order to actualize in medium-long-term basis, the establishment and management of viable science parks in Nigeria decided that the project scope should include: scientific inventory of all relevant technology sources, identification of potential SETI clusters that would serve as technology sources based on location of institutions and their technical competences, identification of potential locations for the parks using aerial imagery technique, processing the location in terms of feasibility and business planning and further designing the required infrastructure locations to enable natural evolution of Science, Technology and Innovation Parks, formation and incorporation of zonal science parks as limited liability companies, managed largely by private sector stakeholders and lastly, specifically studying closely the zonal park located in the Niger Delta region for consideration and immediate implementation.

The Ministry, by September 2007, has done the inventory and location of centres/sources of innovation and technology and surveyed the innovation centres as shown in the Table 21.2.

Based on the above survey, a cluster approach was adopted to ensure viability of the proposed Science Parks *i.e.* each proposed park should benefit from the park facilities in partnership or network basis. Each zonal park operating by harnessing the technology and innovation outputs from the surrounding technology based institutions.

Table 21.2: Innovation Centres in Nigeria

Sl.No.	Technology Sources	*S.E	S.S	S.W	N.W	N.E	N.C	Total
1.	Universities	14	15	28	7	8	16	88
2.	Polytechnics	7	4	12	8	6	9	46
3.	Monotechnics	1	2	2	2	0	1	8
4.	College of Agric	2	2	5	3	2	8	22
5.	College of Education	7	9	11	11	10	13	61
6.	Research Institute	15	11	31	20	7	13	122
7.	Companies/Industries	–	–	–	–	–	–	–
	Total	46	43	89	51	33	85	347

*S.E: South East Zone; N.E: North East Zone; S.S: South South Zone; N.W: North West Zone; S.W: South West Zone; N.C: North Central Zone.

The zonal parks which are to operate as registered Limited Liability Companies are located and to be registered as follows:

Table 21.3: Proposed Science Parks Companies

Sl.No.	Zone	Park Company
1.	N. W	North West Zonal Science and Technology Park Ltd., Sokoto
2.	N. C	North Central Zonal Science and Technology Park Ltd., Makurdi
3.	N. E	North East Zonal Science and Technology Park Ltd., Yola
4.	S. S	South South Zonal Science and Technology Park Ltd., Asaba
5.	S. E	South East Zonal Science and Technology Park Ltd., Aba
6.	S. W	South West Zonal Science and Technology Park Ltd., Osogbo

- Preliminary studies conducted using basic photographs acquired from remote sensing sources to assess the viability especially as the location relates to existing infrastructure and any potential future development was done.
- Interactions held locally and internationally to create awareness, advocate and test the strategies deployed for proposing the options for developing science parks in Nigeria.

5. Way Forward

Implementation roadmap and major responsibilities of stakeholders on the project are as shown in Table 21.4.

6. Funding of the Parks

Each park is estimated to cost N5.0 billion to establish ($1 = N125) and the federal Government has already invested N500 million in 2006–2007 as take-off fund. The Federal and State Governments, the Private sector, individual as well as relevant international agencies are required to bring up the rest. State Governments in particular are to provide the land in strategic areas to meet location specification of the parks.

Table 21.4: Implementation Roadmap and Major Responsibilities

Sl.No.	*Activities*	*Period*	*Stakeholders*
1.	Preliminary Scientific activities to refine the Parks programme implementation	2007–2008	Federal and State Governments
2.	Land allocation and Parks Physical Planning	2007–2008	Ditto
3.	Development of basic infrastructure	2008–2010	Ditto
4.	Establishment of technology based facilities and companies and functional take off of some Parks.	2008–2010	Federal, State Governments and Private sector technology and innovation institution
5.	Full functionality of all parks	2010–2015	All stakeholders

7. Partnership and Responsibilities

Close partnership between major stakeholders *i.e.* the Federal Government, State Governments and the Local Government Area Authorities is envisaged to achieve the objective of successful establishment of Science Parks to drive the economy.

The Federal government, through the Federal Ministry of Science and Technology, Education, Industry Ministries and others is highly committed to ensure that conducive environment prevails to facilitate technology transfer, quality product development, injection of deletion programme on imported technologies and consumable products to strengthen the technology driven local industry.

State Governments are expected to include development of the Parks (in their areas) in their development plans. Budgetary allocation of funds should therefore be made to develop relevant road networks, power supply lines, communication system, railway lines, water supply lines and other infrastructural facilities.

The Organized Private Sector is expected to respond to the opportunities offered by the SPs by investing in development of industries, incubation centres, service and recreational facilities adequately and timely.

8. Discussion

The Raw Materials Research and Development Council (RMRDC) which was established in response to the International Monetary Fund (IMF) imposed Structural Adjustment Programme (SAP) on Nigeria (1986), that made the cost of importation of raw materials prohibitively high, resulting in the close down of industries, came up with the Raw Materials Display Centres. The Centres were designed to stimulate interest in local raw materials development and utilization for the local industries.

Visitors to the Centres *i.e.* investors, entrepreneurs, scientists, Government officials, students of tertiary institutions and the general public were exposed to the raw materials endowment of the country, their locations, reserve estimates, process technology and in some cases models of process equipment to transform the raw materials into their final products.

State Government officials in early days of RMDC establishment; bring in their foreign visitors (investors) to the Centres to have first hand information of the raw materials potentials of the State.

Students of tertiary institutions and even secondary school children visit the Centres as part of their learning programme. Some of them physically seeing for the first time materials they have discussed in the classrooms.

Scientists looking at the Display materials from the raw materials component to the final products were also challenged to find local alternative to raw materials components which were hither-to imported at high cost.

Another activity focused on the Centres apart from students/public excursions in the earlier days of establishment was the Quarterly Display Seminars organized by RMRDC in conjunction with the Organized Private Sector (OPS) like MAN, NACCIMA, and NASSI. Members of these bodies, along with relevant government agencies/parastatals; research institutes, scholars/researchers in universities and the general public used such fora, to discuss and find solutions to problems of raw materials sourcing, development and utilization in a particular manufacturing sector quarterly. Proceedings of such fora were usually published by RMRDC.

Workshop which involved real practical demonstration of production techniques to transform the raw materials into final products were also conducted at the Centres for specialized target groups such as; young school leavers, women groups, cooperatives societies, non-governmental organizations, Church/Mosque groups, producers/suppliers of specific raw materials etc, all in the bid to provide employment opportunities to these groups and in addition add values to the raw materials produced locally.

The change from the original concept of the Raw Materials Display Centres to Raw Materials Information Centres was informed by the realities of prevailing intrusion of ICT into the everyday life of human beings everywhere. The massive growth of the digital technology made it possible for interested parties separated even by Ocean(s) to have access to information which hitherto have to be physically present to obtain or send to them through the conventional postal mails, which could be unreliable, costly or even not timely. This is now made possible online via the creation of ever-more popular websites.

Information can also be made easily and readily available through exchanges with national, regional and international organizations involved in the same business activities. Physical movement of people in quest for information and materials can thus be reduced/restricted to barest minimum if need be.

The linkage of Research Institutes (RIs) in Nigeria with other agencies through VSAT facilities made possible by RMRDC, was first of its kind in the country. For the first time the RIs could link up with themselves and even connect to the outside world to tap into the Information Super way.

On records were other attempts made by Nigeria to steer States/regional and even the national economies to be technologically driven before the bold attempt of the FMS&T to establish Science Parks in 2006.

Worthy of mention are (*i*) Industrial Development Centres (*ii*) Technology Business Incubation Centres (*iii*) Industrial Parks (*iv*) Free Trade Zones and (*v*) the FMS&T advanced Sheda Science and Technology Centre (SHESTCO). These attempts have not largely provided the necessary impetus needed to drive the economy technologically.

The Science Parks as envisaged by FMS&T is expected to provide the launch pad that start-up companies needed when they are "spun-out" from universities or a company (Wikipedia) (2008). Science Parks are also being developed world-wide to leverage the assets of non-university research and development organizations such as Research Institutes (in Nigeria). There are specialized agencies of government with often unique facilities and equipment dedicated to particular areas of endeavor. SPs could therefore provide a location for start-up companies created to commercialize technologies developed in these organizations.

The six (6) regional/zonal SPs, planned for Nigeria by the FMS&T are expected to be located in specialized areas and be serviced by centres of innovations around them. The Organised Private Sector participants *i.e.* MAN, NACCIMA, NASSI, NGO, Individuals, etc are also supposed to be actively involved to realize the vital role these centres could play in the economy.

It is largely the combination of these three interacting elements *i.e.* government, the universities/research institutes and the private sector players that will hopefully give the SPs their dynamism to excel, thus justifying the effort in investing seriously in setting them up.

The State Governments, Local Governments Areas Authorities have been invited to provide the location and infrastructures such as road network, water and electricity supply, railway connection, etc to make the SPs environment conducive.

9. Conclusions

The Raw Materials Display Centre, which metamorphosed into the Raw Materials Information System, is one of the various attempts made by the Federal Government of Nigeria to use science to drive industrialization. Other attempts over time which include: Industrial Development Centres, Technology Incubation Centres, Industrial Parks, Specialized Advanced R&D Complex, Free trade zones etc, have not yielded the much desired goal of driving the economy of the country technologically, hence the recent attempt by FMS&T to establish Science Parks. This attempt to establish SPs in the six geopolitical zones, through a deliberate planning and collaboration between States and Local Government Authorities, Centres of Innovations, the Organized Private Sector participants, and International collaborators etc must therefore be commended and supported by all stakeholders. In realizing this objective, care must be taken not to use in to-to strategies used by other countries in the establishment of their SPs, without taking cognizance of the existing "realities on ground". It is essential to point out that the establishments of the Parks are closely linked up with the aspiration of the nation of being among one of the 20 leading economies in the world by 2010.

Existing facilities in Government agencies *i.e.* the various Ministries, Parastatals, Innovation Centres and the OPS must be properly harnessed to strengthen the proposed parks. The RMIC programme of the RMRDC and its facilities are thus expected to dovetail into the zonal SPs. Officials of the Council are prominent members of the Standing Committees inaugurated by FMS&T for the establishment of SPs in the country.

Visionary personalities from each zone of the country are expected to champion the zonal park companies by investing their needed political clout and financial resources to speed up the implementation of the project. The National and State Assemblies are expected to give the necessary support by passing the requisite bills to enable the executive Arm of the Government implement the programme.

Detailed study of the innovation systems of the country still needs to be done to technologically build the knowledge base required to energize the project and assist in refining the proposal and determine the functional and technological sensitivities of the proposed parks. Nigeria experts in Diaspora must also be involved in this programme.

Deliberate effort must still be made to develop the physical infrastructures required to service the parks. The erratic and epileptic power supply system, poor road network, railway system etc must be improved upon.

The innovation centres must be given necessary impetus to "clean up" their acts, be more responsive and come up with knife-edge, market driven technologies/ innovations, that would readily be taken up by investors.

The OPS must be challenged, energized and supported financially. The momentum created by the recent (reforms) recapitalization exercise in the banking sector must be sustained in order to pool together funds trapped in the informal sector of the economy to fund industrialization. The vibrant macro-economic climate engendered by relatively calm political situation in the country, the recent Public Service Reforms, the removal of the debt burden, the burgeoning foreign exchange built-up, are all pointers to the fact, that the country is poised/primed for economic transformation. Creation of Science Park is therefore one of the planks to be used by the Government to launch Nigeria into the knowledge based 21st century. The country therefore needs all the support, *i.e.* technical and financial of international collaborators to establish the Science Parks and make it work in Nigeria.

Acknowledgements

My sincere appreciation goes to the organizers of this seminar, in particular M. Bandyopadhyay and A. P. Kulshreshtha for accepting me to present this paper despite the initial challenges. I will also want to put on record the financial support of the Centre for Science and Technology of Non-Aligned and other Developing Countries.

My appreciation also goes to the Honourable Minister of Science and Technology for the approval to attend this seminar and the use of some of the technical papers of the Ministry. Same also goes to I. O. Adegoke, T.N. Onyemah, N. Nwaokoro, D. B. Ayo and A. P. Onwualu and the entire Management and staff of RMRDC for the partial financial support and the assistance rendered in the preparation of the paper.

References

Federal Ministry of Science and Technology, 2007. The Development of Viable Science Parks in Nigeria.

Raw Materials Research and Development Council, 1991. Investors Guide to the Raw Materials Resources of Ogun State, Nigeria.

Thompson, E., 2007. Raw Materials Information System Paper presented at 2-day Summit on the state of Raw Materials Research and Development in Nigeria.

Wikipedia (2008); Science Park. http:/en.wikipedia.org/wiki/Science Park (accessed 1st Feb. 2008)

Abbreviations

- RMRDC: Raw Materials Research and Development Council
- RMDC: Raw Materials Display Centre
- RMIC: Raw Materials Information Centre
- RMIS: Raw Materials Information System
- FMS&T: Federal Ministry of Science and Technology
- UNDP: United Nations Development Programme
- V-SAT: Very Small Aperture Terminal
- WAN: Wide Area Network
- PSnet: Public Service Network
- SETI: Science Engineering, Technology Innovation
- OPS: Organized Private Sector
- MAN: Manufacturers Association of Nigeria
- NACCIMA: National Association of Chambers of Commerce, Industry Mines and Agriculture
- NASSI: National Association of Small Scale Industries

Chapter 22
Science Culture in India

Vinay B. Kamble
Vigyan Prasar, Department of Science and Technology, Government of India
A-50, Institutional Area, Phase II, Sector 62
Noida – 201 307, Uttar Pradesh, India
E-mail: vbkamble@vigyanprasar.gov.in

ABSTRACT

Science centres and science museums are the nuclei where begins a process of transformation of a country into a nation of scientifically minded people. The presentation deals with the development and progress that has taken place in taking science to the people through the medium of science centres and museums with emphasis to India, and the change that is being perceived. In particular, it emphasizes the fact that the new role science centres and museums need to play is to act as a bridge between the people and the R&D (Research and Development) laboratories from where the new technologies flow; and explain the impact of science on environment and society; and, help people make informed choices.

Keywords: *Science Centres, Peoples science, Science museums, Science communication, Public understanding of science, Vigyan prasar.*

1. Introduction

Science communication in India has its roots in the scientific renaissance in the late nineteenth century in West Bengal and owed it to the efforts of Mahendra Lal Sarkar, Fr. Eugene LaFont, P. C. Ray, Ashutosh Mukherjee, and Jagdis Chandra Bose through the establishment of the Indian Association for cultivation of Science. Besides encouraging scientific research, the Association put in efforts to take science to the people through public lectures and exhibitions. It was also around this time that Ruchiram Sahni initiated a movement to take science to the people in Punjab. People in Punjab would throng to listen to Ruchiram Sahni's lectures by paying a fee of two annas. One anna, incidentally, represented one-sixteenth of a Rupee! Also, in the late nineteenth century, Madhavrao Namjoshi organized an Industrial Exhibition in Pune with an aim to give an impetus to the industrial development in India. Indeed, people do possess a keen desire to know and understand the new science and new technologies as they become part of their lives. The innate curiosity and the thirst for knowledge always co-exist. Otherwise, over a century ago, people in Punjab would not have thronged to listen to Ruchiram Sahni's lectures by paying a fee of two annas. In the present century, exhibitions on Science and Technology (S&T) have become an integral part of annual sessions of the Indian Science Congress.

With the impact of Science and Technology (S&T) on social life and economic development, the importance of S&T popularization was realized decades ago in India. This concern is reflected in the scientific policy resolution of 1958 and 2003; and even in the Constitution of India. Notable efforts in Kerala in taking science to the people in 1960s culminated in the establishment of Kerala Sastra Sahitya Parishat (KSSP). It is important to note that KSSP was instrumental in pioneering a people's science movement in India.

With establishment of governmental agencies like National Council for Science and Technology Communication (NCSTC) and Vigyan Prasar (*vigyan* stands for science in Sanskrit or Hindi, and *prasar* stands for spread or dissemination), and other voluntary agencies coming into being in the last twenty five years or so, a variety of activities now take place all over the country. National Council for Educational Research and Training (NCERT) and National Council of Science Museums (NCSM) have been contributing to the field of S&T popularisation significantly. A large network of centres for awareness about modern agricultural practices–also called Krishi Vigyan Kendras–under the Ministry of Agriculture serves the farmers and their needs.

2. Science Communication in India: Present Scenario

NCERT has a mandate of developing curricula and preparing text books at school levels, training and research in the field of education. NCERT also organises programmes like science exhibitions for schools. National Council of Science Museums (NCSM) is the largest network of science centres/museums in the country. Besides, there are science centres such as the Homi Bhabha Centre for Science Education, Mumbai, Vikram A. Sarabhai Community Science Centre, Ahmedabad, and Vidnyan Ashram in Pune, engaged in non-formal science education and popularisation of S&T. There are 24 state-of-the-art planetaria spread in different parts of the country, and zoological parks.

S&T coverage in the newspapers/magazines is also steadily picking up. Over the last five years, most newspapers in the country carry a science page/supplement at least once a week. This holds true not only for English and Hindi newspapers, but also for the regional newspapers in other languages. Further, popular science magazines have proliferated in several regional languages. As regards the other traditional media, All India Radio (AIR) broadcasts at least one programme everyday on a scientific topic from all its stations, in addition to the ones on health and agriculture. Several AIR stations with science cells broadcast three programmes per day. Doordarshan–the National Television Channel–telecasts about two programmes every week on the national network and all regional centres put together produce and telecast some 150 S&T programmes every year.

Agencies like University Grants Commission, Central Institute of Educational Technology, Indira Gandhi National Open University, and Vigyan Prasar also have regular slots on Doordarshan. Programmes on S&T Popularization for telecast/ broadcast are also produced by several Government/Non-Government agencies.

National Children's Science Congress is being organised by NCSTC for the last fifteen years in the country and has already caught the imagination of a few Western and West Asian countries. In this programme some 600,000 school children participate. They undertake studies/projects on specific scientific themes. The reports are then submitted at school, district, state and national levels. The winners then participate in the special session of the annual Indian Science Congress in January every year. School children from a few other countries participate in this event on a regular basis.

Needless to say, National Children's' Science Congress, S&T popularisation activities built around the total solar eclipses of October 24, 1995 and August 11, 1999, Venus transit of 08 June 2004, World Year of Physics 2005, Emergence of Modern Science, S&T for the visually challenged, teaching/learning aids and toys, networking of Government/Non-Government organisations for S&T communication, are all pioneering efforts in S&T Communication. Bharat Jana Vigyan Jatha 1987 and Bharat Jan Gyan Vigyan Jatha 1992 (groups of science communicators) moving in different parts of India, spreading science awareness, interacted directly with people from nearly 400 districts covering about a fourth of India's population.

Mention also may be made about the Vigyan Prasar Network (VIPNET) of Science Clubs–mostly in rural areas, with nearly 8,000 member clubs as of today. This has laid the foundations of a national science club movement. A countrywide network of Eco-clubs established to spread awareness about conservation of environment, biodiversity and sustainable development has also been established by the Ministry of Environment and Forests.

A unique effort was undertaken to take science to the people in a train called the "Vigyan Rail" or Science Exhibition on Wheels that moved throughout India for about eight months from December 2003 to August 2004. During its run, it travelled to 60 destinations covering 15,000 kilometres, the number of visitors being nearly 5 million. It had working models, interactive exhibits, computer games and quizzes, charts, scientific videos and so on. The exhibition was the result of cooperation and collaboration between 18 Ministries/Departments of Government of India engaged in various scientific fields.

Incidentally, constituent units of NCSM, science centres, R&D laboratories and a number of Government and non-Government agencies participated in this gigantic effort.

3. Science Centres in India: A Few Milestones

After Independence, there were a few examples of establishment of science museums through individual initiatives. Notably among them was the establishment of the Arthropoda Museum in Pune–the very first natural history museum of India housing a large collection of crustaceans, insects, lobsters, prawns, crabs, beetles, worms and cockroaches–by S. M. Ketkar in 1961.The museum conducts educational and research activities, biological competitions for Zoology students, nature study tours and short tem courses in entomology; and has become a tourists' attraction in Pune. Much later, the National Museum of Natural History was set up in 1978 at New Delhi.

Community Science Centre established at Ahmedabad in 1966 by the visionary scientist Dr. Vikram A. Sarabhai, the Father of the Indian space programme, is engaged in the non-formal science education and science popularization among the schools and the general public. It has been the fountainhead of inspiration for establishment of science centres in different parts of the country. It is interesting to note that Science Playground is a unique concept that travelled from the tiny Community Science Centre to the West through National Council of Science Museums (NCSM). Of late, Exploratory Science Centre has come up at Pune with similar objectives. It conducts enrichment activities for school children and engages them investigative projects. In the last five years or so, the concept of science parks has even penetrated the schools. It is not uncommon to find science parks in schools today, especially in urban areas.

NCSM is the largest network of science centres/museums in the country under a single administrative umbrella in the world. 27 science centres/museums of NCSM located in different states in India and are visited by around 6.5 million people every year.

Establishment of Science City at Kolkata by NCSM was a milestone indeed in the history of S&T communication in India. Spread over a huge area, it offers a journey through the Jurassic period with different species of dinosaurs, a popular science hall, large format theatre showing 3-D scientific movies, and a science park. Over the years, two science cities have come up at Ahmedabad in Gujarat and Kapurthala in Punjab.

Vigyan Prasar, under the Ministry of Science and Technology, has flourished into a national resource-cum-facility centre; and is engaged in the development of a variety of software utilizing different means, media and modes. Vigyan Prasar has been regularly beaming weekly television programmes aimed at communicating science to the people for over three years now. Vigyan Prasar also has been producing radio programmes on various aspects of science and technology and broadcasting on All India Radio in major Indian languages. At present, a 56-episode radio serial on Planet Earth is on the air and is simultaneously broadcast in 19 major Indian languages from over 100 All India Radio stations in the country. Vigyan Prasar has set up a network of satellite interactive terminals spread throughout the country exclusively for S&T communication with two way audio-two way video facility using Edusat, India's satellite for education. Further, since the last four years, popular science programmes are broadcast daily for one hour on WorldSpace radio. An analogue to digital interface developed for PCs by Vigyan Prasar allows measurement and control of analogue variables like temperature, intensity of sound and light, pH, and so on. NCERT and school establishments have evinced a keen interest to introduce these experiments in the school curriculum. Yet another important activity of Vigyan Prasar is the utilization of the print media and the Internet for S&T popularization.

4. Public Understanding of Science–India

The India Science Report 2005 (ISR) presents the state of science and technology in India. Incidentally, this is the first quantitative report that also gives an insight into the public understanding of science. In particular, it states that there is no decline in interest in the proportion of students who wish to study science. But, a third of the students said they did not study science as they did not feel motivated enough and another 40 per cent said the number of students in a class were too many for them to understand what was being taught! On the other hand, half the teachers interviewed believed that more computers/equipment were required for teaching science subjects since inadequate science training was a serious issue. Since every generation of top quality scientific manpower starts at the school level, a lot also depends on the way science is taught at school levels. Surely, this is an area where we need to focus our attention.

As regards the Human Resource in S&T (HRST), ISR states that India has 52.6 million graduates, post graduates and diploma holders. If we remove 12.2 million unemployed and housewives from this category, we get a total of 40.2 million that form an S&T resource base. Those who have a diploma/graduation degree and are employed in a science and technology occupation comprising the HRST 'core' group are around 14.2 million. ISR 2005 draws interesting inferences as regards public attitude towards S&T.

Over three fourths of the public feel that S&T is important for education; and believe that S&T makes lives healthier and more comfortable. On an average, the level of knowledge the population has about the scientific concepts is very high–57 per cent of the people knew that the centre of the earth is hot and 86 per cent knew that that the oxygen we breathe comes from the plants. Not surprisingly, given how women are blamed for not having a male child, just 38 per cent knew that the sex of the child depends upon the father. Surely, the answers to science related questions tend to be increasingly correct as the education levels of the respondents rise.

What are the sources of information of the public? Television remains the primary source for 57 per cent of the people of the country, and is almost five times more popular than the newspapers. Close to three-fourths of urban households rely on TV for information, as do half the rural households. Indeed, even educated people rely more on TV than on any other medium. In the case of postgraduates, 65 per cent rely on TV as the primary source of information compared to just 27 per cent on newspapers. Close to two-thirds of the population gets its science related information from TV as compared to 8 per cent from newspapers. Over three-fourths of the people (85 per cent) have a great deal of confidence in the authenticity of the TV, and ironically it is the illiterate that have the least confidence (64 per cent). Nearly 65 per cent of S&T news is got from TV in India, as compared to 7 per cent in the US. The report finds that television is the most popular source of information for most people. But this also calls for a conscious action on the part of all concerned to generate quality S&T programmes for television. Quality S&T TV programmes are few and far between. There is no gainsaying the fact that this source of dissemination of scientific information needs to be exploited fully. This finding makes a strong case to utilize television and the Edusat infrastructure for S&T communication.

How about Internet as a source of information? Internet, however, does not appear to be popular source of information in India. Over 44 per cent of S&T information in the US is obtained from the Internet as compared to 0.2 per cent in India at present! There is a need to ensure greater penetrability of Internet and other ICT tools at the school level as also in rural and remote areas so that access to reliable and updated information is considerably improved.

But what is of prime concern is the extremely low percentage of people visiting science museums, planetaria, aquaria, science fairs, etc. Is this due to less awareness or less motivation? This needs to be ascertained, for, these are places that document scientific and technological developments and could be a great source of scientific information as well as inspiration for children. While 32 per cent of Indians have never visited a cinema hall or seen a video, the figure is around two-thirds for places such as a science institute/park/museums/planetarium and zoos/aquariums. Under three per cent of the families (5.7 per cent in urban areas and 1.7 per cent in rural ones) have visited a science institute once while around 12 per cent have visited a museum once, a fifth have been to a zoo, and just eight per cent to a library. Indeed, these are the stark realities on which the future strategies of science centres and museums would need to be based.

5. Ground Realities

In view of the fact that our world is largely shaped by science, it is assumed that everyone needs to understand science. Surely, many science related topics–say, construction of a dam, construction of a nuclear power plant, or introduction of genetically modified crop–do have implications as regards public policy. There could be a long list of such highly important issues. But, quite often, in many science centres one sees an over-emphasis on fun and entertainment. One reason could be to attract children forming the largest group of visitors to the science centre. Another reason could be the revenue generation on which heavily depend many science centres.

However, in the process, important scientific issues like global warming and climate change are lost sight of. It is equally true that in bigger cities, science centres compete with a host of other options of entertainment and fun. This could be one reason why science centres often shy away from complex issues the society faces, but it deprives the adult audiences from learning about issues of relevance. It is here where the science centres need to re-invent themselves.

Science Centres often present static exhibits and dioramas giving drab information which visitors may or may not find interesting or relevant. Surely, such exhibits hardly ever generate any interest among visitors. The result is that few visitors pay a second visit to the science centre. In order that a science centre continues to attract visitors time and again, it must present exhibits, live events such as lectures/ demonstrations, video, animations, digital updates, and live links to scientists working in specific fields. Further, interactive exhibits provide hands-on experience, help generate interest and understand a scientific concept in an effective manner. Development of thematic galleries on current topics, cutting-edge technologies, and issues like nanotechnology, global warming and climate change, need to become a continuous process thereby helping attract visitors more often to the science centre.

6. Issues that Need to be Addressed

Literacy in India has significantly gone up over the years. Nearly 65 per cent of the people can read and write today. Higher literacy rate along with liberalization of economy have raised the expectation of the people from the science centres. We are on the brink of exciting developments in science that will affect everyone's lives. Along with the benefits development of new technologies come potential risks which cause concern. People rightly expect proper safeguards against risks, and full and open discussion about the values, benefits and risks of scientific developments.

New technologies are fast replacing the "what-were-once-new" technologies as they become obsolete. How fast the "latest" model of a computer purchased by you becomes obsolete, or how fast compact fluorescent lamps are replacing the good old tube lights. In the field of medicine, terms like ultra-sonography, magnetic resonance imaging (MRI), organ transplant, and in-vitro fertilization have almost become a part of everyday life. Consider the advances in the field of biotechnology. Genetically modified food products, transgenic crops, cloning, gene replacement therapy, stem cell harvesting and similar jargon is fast becoming a part of our everyday vocabulary. We are faced with issues like the global warming and climate change; and the issue is further aggravated by the pumping of greenhouse gases as a result of anthropogenic activities. Science Centres must not shy away from addressing these issues that have a profound social relevance.

Undoubtedly, the new technologies have helped improve the quality of life, thereby significantly changing our life-styles. However, it is also equally true that breakthroughs in the fields like biotechnology have instilled a sense of awe and fear–fear of the unknown–in our minds. Consider the question–are genetically modified food products safe to consume? Will transgenic plants with built-in pesticides give rise to a new breed of pests which are resistant to these built-in pesticides? Most of us are totally ignorant of what the new science holds in store for us.

There has been a marked change in the social fabric over the last decade as a combined result of the liberalization of the economy and the phenomenal growth in the field of information technology. This process of development has its own associated problems. More and more students have been opting for professional courses and fewer and fewer for a career in pure science, which could result into an acute shortage of scientific manpower in near future. This imbalance will need to be corrected, since science and technology need to go hand in hand for an all round growth of a society. It is here that science centres and museums can make a difference. Active participation of science centres and museums contiguous with programmes like the National Children's Science Congress, science exhibitions, Emergence of

Modern Science, contact programmes for school and college students with scientists, and visits to R&D laboratories, would go a long way in explaining to them the thrill and challenge offered by a career in pure science.

There is no gainsaying the fact that the different sections of society that would need to be specifically addressed include school children at primary, secondary and tertiary levels, out of school children, women with emphasis on girl-child and housewives, adult illiterates, employees/workers/artisans, farmers and agriculturists, and college/university students. Indeed, it is a mammoth task to address all the groups, since the software and efforts required for different groups listed above would need to be tailor-made. This would require intensive research in determining their needs and expectations–both in urban and rural areas. It is then that we can expect a higher percentage of people to visit science centres and museums seeking answers to their questions.

7. Science Centres–Looking Ahead

The fact that science centres are educational institutions where education takes place through informal learning needs no over-emphasis. Science centres have an important role to play in preparing teachers and parents to teach science. Science centres provide a first hand experience to people of all age groups in a rich learning environment that is free from the constraints of the formal system.

In view of what India Science Report 2005 says on public understanding of science, what should be the role and responsibilities of science centres and museums? For the development and growth of a society, the spirit of enquiry and scientific attitude need to be inculcated at every level in every system. There is no gainsaying the fact that spirit of enquiry does not relate only to scientific matters, but, it equally permeates other spheres of human activity as well, say, language, society, sports, administration and so on. Indeed, the aim of a science centre or a museum should be to disseminate scientific and technological information to the people on various aspects and issues that directly affect their lives and to inculcate a spirit of enquiry among them in order that they can make informed choices.

Surely, the major function of science centres and museums is to supplement the formal system of education as imparted in schools and colleges. One encounters science in every walk of life, say, in agriculture, in medicine, in nursing, in child care, in mitigating disasters and so on. It is important to realize how various disciplines of science come together in an integrated manner when we observe a natural phenomenon. Even in social context, while dealing with problems and issues related to health, environment, water, energy, and so on, one needs to consider physical, chemical and biological aspects simultaneously. It is hence imperative that a holistic and integrated view of the phenomena we observe in our environment–both physical and social–is given. This will help people appreciate the intimate relationship of science with their lives. It is in this sense that science centres and museums are Facility-cum-Resource Centres.

Over the years, the distinction between a science centre and a museum is slowly getting blurred in terms of activities. With growth in literacy in the developing countries, changing socioeconomic scenario, and the revolution in the communication and information technology have made science centres and the science museums to redefine their roles. A museum is no longer a place with a mere collection of items and exhibits, or a science centre with a few activities for children. Their prime responsibility is to make people scientifically literate, trigger an interest in science among them– especially the young, and make them understand that science is everywhere and for everyone. The primary role the science centres and museums must assume today is to act as a bridge between the people and the R&D labs from where new technologies flow and explain their possible impact on the environment and society.

Conclusions

We have briefly discussed the development and progress that has taken place in taking science to the people through the medium of science centres and science museums with emphasis to India, and the change that is perceived. In particular, it is emphasized that the extremely low percentage of people visiting science centres, museums, planetaria, aquaria etc. especially in India, is of grave concern as regards dissemination of scientific information and that science centres need to redefine their role and priorities. In particular, the need to address scientific/technological issues that have a direct impact on environment and society has been brought out. It is stressed that the primary role of science centres and museums is to act as a bridge between the people and the R&D labs.

References

Padmanabhan, B. S., 2006. Vigyan Rail–Science Exhibition on Wheels.

Sehgal, N. K. and Mahanti, S. (Eds.), 1994. Memoirs of Ruchiram Sahni.

Shukla, Rajesh, 2005. India Science Report, National Council of Applied Economic Research.

Part IV

Science Centre Policies

Chapter 23

State Support for the Development of a Network of Science Centres in South Africa

K.M. (Lebs) Mphahlele* and T.I. Ramovha**

Department of Science and Technology (DST), South Africa

*E-mail: *komane.mphahlele@palama.gov.za, **isaac.ramovha@dst.gov.za*

ABSTRACT

In March 2007, Cabinet approved the Youth into Science Strategy whose main strategic objectives are to: (a) promote science and technology literacy, and (b) identify and nurture talent and potential for science, engineering and technology-based careers. It also approved the establishment of a Network of Science Centres that will serve as an infrastructure to deliver the Youth into Science Strategy. The establishment of a Network of Science Centres is guided by a 25-year National Rollout Plan which spells out two models of implementation: *viz.* the redress model and the public-private partnership (PPP) model. The former advocates for the development of what the Norms and Standards for a Network of Science Centres refers to as a 'limited service science centre' while the latter advocates for the establishment of what is referred to as a 'full service science centre'. The role of Government will be to support the implementation of the redress model while private sector will assume substantial financial, technical and operational risks in the design, financing, building and operation of the PPP model. This paper presents progress made by the DST in creating a policy environment enabling local science centres to fulfill their role with regard to the establishment of a Network of Science Centres as approved by Cabinet.

Keywords: *Science centres, Network of science centres, Science and technology literacy, Youth into science strategy.*

1. Introduction and Background

The South Africa's watershed democratic elections of 1994 ushered in a new political dispensation which is underpinned by principles of freedom, non-discrimination, redress, access and equity. On the education front, the various race and ethnic-based Education Departments were amalgamated into a single non-racial, non-sexist Department of Education. In the first five years of democratic rule, the transformation of the education system was accompanied by the unintended decline in learner participation and performance in, especially, mathematics and science.

In order to arrest this down-ward trend, the Ministry of Education launched the National Strategy for Mathematics, Science and Technology Education in 2001. The results of this national effort have been encouraging (Table 23.1).

Table 23.1: Performance in Mathematics and Science in the Last Ten Years

Year	*Total Candidates Passing*	*Maths (Higher Grade)*	*Maths (Std Grade)*	*Science (Higher Grade)*	*Science (Std Grade)*
1995	283 742	29 475	54 947	34 895	28 020
1997	264 795	19 575	65 580	26 658	34 584
1999	249 831	19 327	79 631	23 344	54 884
2001	277 206	19 504	72 301	24 280	45 314
2003	322 492	23 412	99 426	26 067	61 756
2004	330 717	24 143	109 664	26 975	73 943
2005	347 184	26 383	112 279	29 965	73 667
2006	351 217	25 217	110 452	29 781	81 151

Although participation and performance by the historically disadvantaged in school Mathematics and Science improved, it still falls short of being representative of country's demographics. In higher education, disproportionate low enrolments and success by Blacks and Women in scarce skills continues (Tables 23.2 and Figures 23.1 and 23.2).

Table 23.2: Performance in Scarce Skills Disciplines According to Race

CESM	*Black*	*White*
04: Business/Management	33 per cent	72 per cent
15: Life and Physical Sciences	31 per cent	63 per cent
16: Mathematical Sciences	35 per cent	63 per cent
22: Social Sciences	34 per cent	68 per cent
12: Languages	32 per cent	68 per cent

In addition to low and un-representative rates of participation and performance in mathematics and science education, South Africa is characterized by acute 'engineering gap'–*i.e.* shortage of engineers due to low enrolment rates in engineering fields, low throughput rates, increasing number of unregistered engineers, increased

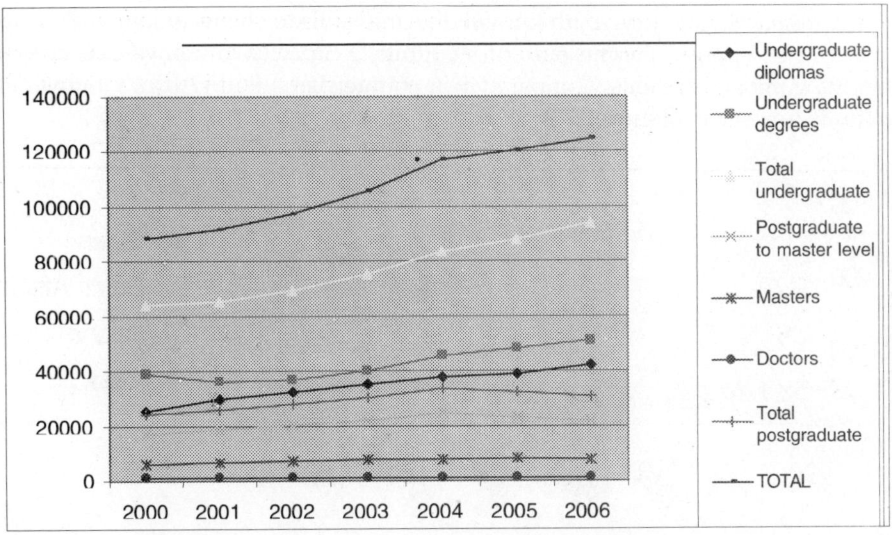

Figure 23.1: Graduation Rates According to Level of Study

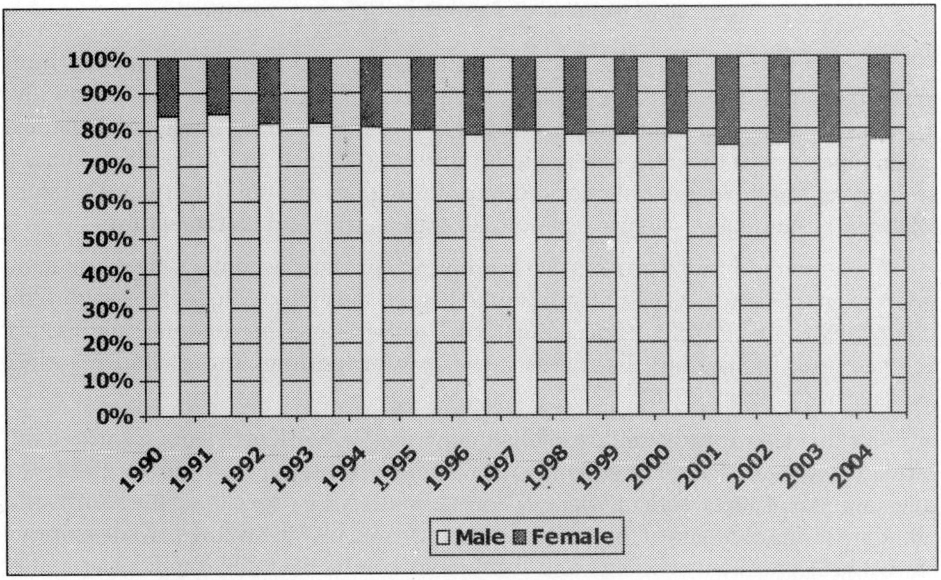

Figure 23.2: Skewed Gender Participation in Higher Education

demand for engineers and emigration of engineering professionals. The capacity of higher education institutions to drive and lead development of skills to innovate is also stretched. When compared to other emerging economies such as Brazil, Taiwan and South Korea the state of knowledge production in South Africa is poor.

For South Africa to succeed in a rapidly changing world it would require developing the skills of its people to the fullest possible extent, carry out world class scientific research and apply both knowledge and skills to create an innovative and competitive economy. The measure of a country's capacity to innovate is, among others, the number of patents. Compared to its counterparts, South Africa's innovation capacity is in a stasis (Figure 23.3).

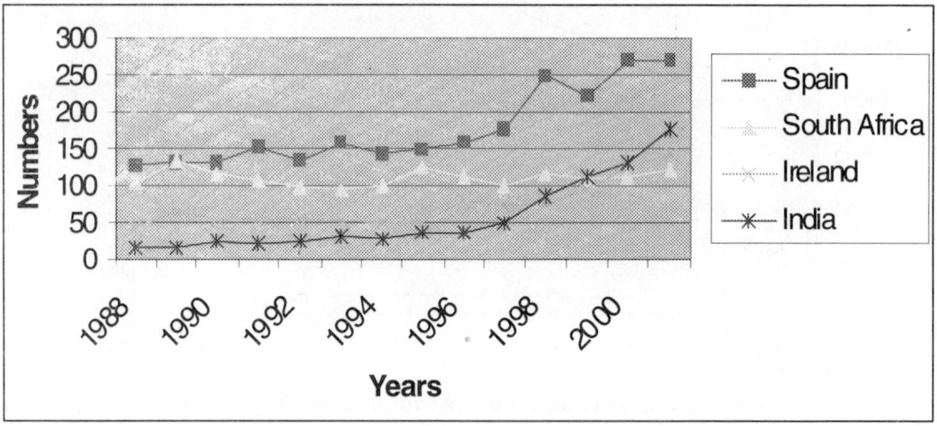

Figure 23.3: Number of Patents in Different Countries

In order to address the above challenges of inadequate or sub-optimal participation and performance rates in mathematics and science as well as low innovative capacity, Cabinet approved the Youth into Science Strategy in 2008. The main objectives of this Strategy are to (*a*) promote science and technology literacy among the public in general and youth in particular; and (*b*) identify and nurture talent and potential in science, technology, engineering and mathematics.

Government-funded projects that promote science and technology literacy include the National Science Week (NSW), Public Understanding of Biotechnology (PUB), Platform Months (PMnths), Space Week, SET Awards and International Innovation, Science and Technology Exhibition (these interventions and varied resource allocations in Figure 23.4).

Cabinet also approved the establishment of a Network of Science Centres as an infrastructure to deliver the Youth into Science Strategy. In order to guide the development of a Network of Science Centres, South Africa developed the Norms and Standards. The contents of this policy were informed by the findings of a descriptive study the Department of Science and Technology conducted in 1999 and the feasibility study it conducted in 2004. The findings of the two reports highlighted the unevenness of science centres in South Africa and the need to not only make them more accessible to the majority of South Africans but to also improve their capacity to implement the Cabinet-approved Government's Youth into Science Strategy. To address this need, an intensive consultative process through ten workshops in all the nine provinces produced a draft Norms and Standards document that was supported at a national consultative conference. Inputs on the adopted Norms and Standards for a Network of Science Centres were also received from the International Workshop on 'Changing Roles of Science Centres and Museums in Developing Countries' that was held in

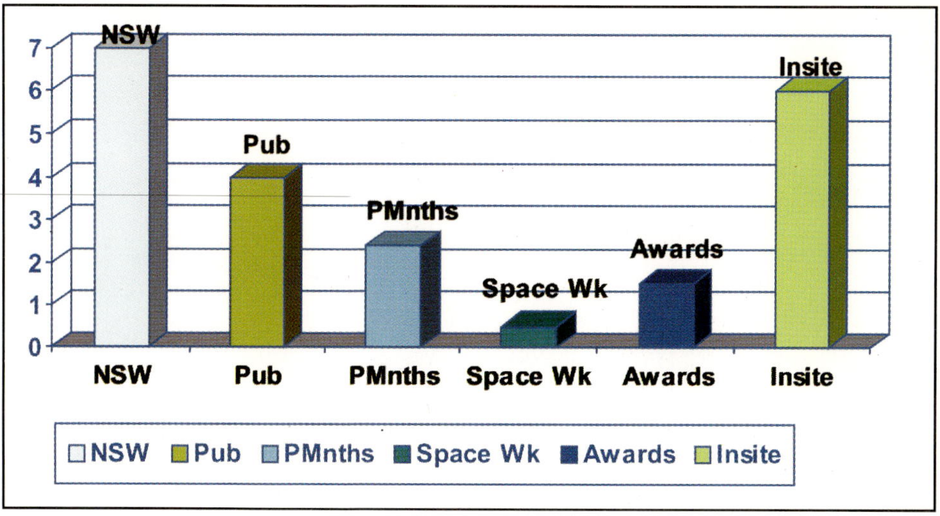

Figure 23.4

Hanoi, Vietnam, in 2004. The Department of Science and Technology approved the Norms and Standards in February 2005.

The approved National Norms and Standards for a Network of Science Centres framework specifies the following goals of the envisaged Network:

- To identify, nurture youth talent and potential in STEM;
- To promote science literacy among the youth and the population in general;
- To enhance learner participation and performance STEM.
- To provide career education in general and STEM in particular to the youth.

The Norms and Standards policy framework also foregrounds the following concepts that are part of the discourse centred on the establishment of a Network of Science Centres in South Africa.

- *Science centre* is a permanently established education facility that provides an interactive educational experience through the use of interactive science, technology, engineering and mathematics exhibits, displays and programmes.
- *Network of science centres* refers to a group of science centres that are interconnected, aligned to and supported by the Department of Science and Technology (DST).
- *Interactivity in a science centre* refers to actions that elicit responses that ideally should be manual, mental, emotional and/or social.

2. Principles

Given the historical backlogs and disparities mentioned by the various studies conducted in South Africa, the Norms and Standards for a Network of Science Centres stipulates that programmes and activities to be undertaken by science centres will be underpinned by the following principles:

- Access, redress and equity
- Key performance indicators-driven delivery
- Outcomes-based Education
- Valuing Indigenous Knowledge Systems
- Sustainability

3. Interim Support for Existing Science Centres

While developing a National Roll-out Plan for a Network of Science Centres in South Africa, the Department of Science and Technology has since 2003 been providing existing science centres with programmatic support as an interim arrangement. The number of science centres that have been benefiting from Government's support has been increasing annually. During the 2003/2004–2008/2009 period financial years, the number of science centres that benefited from the Government's programmatic support increased from eight to sixteen (Figure 23 5 map of these science centres

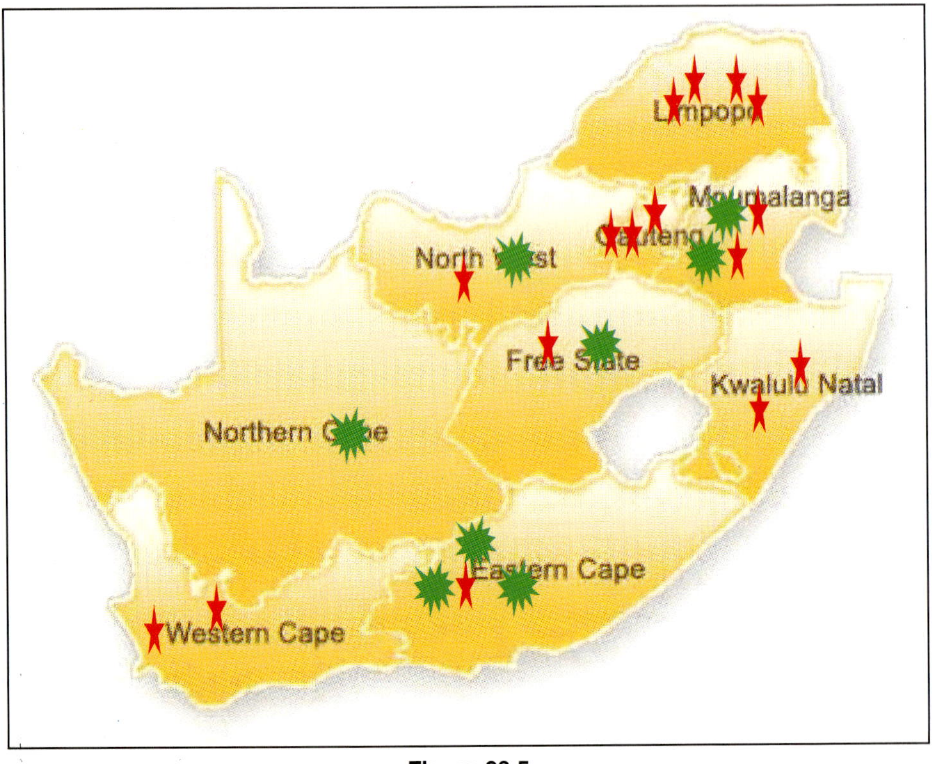

Figure 23.5

(a) The map is not drawn to scale; (b) The red icons represent existing science centres supported by the Department of Science and Technology; (c) the green icons represent planned new science centres.

Thus far, programmatic support to science centres ensured not only the survival of existing science centres but also increased their profile and capacity to implement some of the Youth into Science Strategy interventions. It is envisaged that this form of interim support will be replaced by a more systematic operations support that will be competently managed and executed.

Plans are afoot to formalise and strengthen the collaboration between the Department of Science and Technology and the Southern African Association for Science and Technology Centres (SAASTEC). It is envisaged that such a strengthened collaboration will contribute towards improving the capacity of science centres to implement the programmes and projects that are part of the Youth into Science Strategy. In addition, through its annual conferences, SAASTEC will be able to promote coordination and synergy among science centres in the SADC (Southern African Development Community) region.

4. Long-Term Support for Science Centres

Science centres in South Africa are at various level of development. Whilst some are better resourced, none is in a position to adequately respond to the goals of the envisaged network of science centres in South Africa. This, in addition to shortcomings identified through studies conducted by the Department of Science and Technology, led to a two-pronged approach to enhance the capacity of science centres to respond to the goals of the network which comprised training and programme interventions.

Training Interventions

A training programme comprising nine interventions has been designed to benefit personnel in existing science centres. Implementation of this training programme, which commenced in December 2005, comprises two broad categories, *viz.* the none-core and core areas. The following none-core areas refer to those interventions related to operational management of science centres:

- Financial management course for non-financial managers
- Project proposal writing
- Networking
- Marketing
- Project management

The core areas refer to interventions that relate directly to the core-business of a science centre such as interactive learning, exhibits and displays. To equip science centres with relevant skills to run these interactive programmes, the Department of Science and Technology designed the following core interventions:

Conceptualizing a Science Exhibit and Writing a Brief for the Designer

Educating the audience through interactive exhibits is the major activity of science centres. Science centres should be able to conceptualize their own exhibits and also develop appropriate specifications for exhibit designers and builders. This will reduce dependency on imported exhibits that are expensive to acquire and maintain. It also creates a pool of knowledge to be shared by the science centre community.

Developing New, Interactive Science Programmes Based on Current Issues

More often, people are confronted with natural occurrences they have little or no understanding about. Science centres should be able to respond rapidly to these events by developing interactive awareness programmes to help people have a better understanding of current science issues.

Communicating Science Exhibits to Audience

The success of science centres will, among others, depend on effective communication to the target audience. Equipping science centres staff with appropriate communication skills in this area is necessary.

Impact Evaluation

The studies conducted by the Department identified the conductive of impact evaluation of science centres as an area that needs more attention. Skills to measure impact of a science centre needs to be built over a period and should be part and parcel of a capacity building programme.

Programme Interventions

On the other hand, science centres' performance can only be measured by the extent to which their programmes receive high ratings from their visitors. The Policy Framework for a Network of Science Centres in South Africa has acknowledged the need for science centres to develop and implement programmes in response to the four goals of the Network. As a result, attempts are being made to identify existing programmes that are relevant to the goals of the network such as National Science Week, competitions, olympiads, camps and supplementary tuition. Preference is being given to programmes initiated and/or supported by other Government Departments and/or Government agencies. This is a way of maximizing utilization of useful existing infrastructure and strengthening of cooperative governance.

5. Elements of the Rollout Plan

The Department approved the 25-year National Rollout Plan for the Establishment of a Network of Science Centres in 2006. According to this Plan, a science centre is required to meet the defined admission criteria if it is to be part of the Network. This aspect of the Plan entails the establishment of the accreditation process and the structure that manages the Network. Plans are afoot to finalise admission criteria. This will enable the current loose arrangement of relationships between science centres and Government to be replaced by a more systems approach that is based on approved Norms and Standards.

The National Rollout Plan deals with the following aspects of implementation:

Infrastructure Development

This involves upgrading of existing science centre infrastructure and establishment of new science centres. Under-utilized infrastructure belonging to Provincial Governments will be targeted for upgrading or development into science centres.

Programmatic Support

Science centres and/or the Department will identify programmes to be fully funded by the Department. Other government Departments are also encouraged to use science centres in the same fashion.

Operations Support

The following are being considered to help sustain operations of science centres:

- Internship and learning[1] programmes;
- Fixed rate subsidy for every learner visiting science centres;
- Negotiating for funded posts by Provincial Departments of Education; and
- Fully funded programmes alluded to under programmatic support

Capacity Building

The skills of science centres' personnel remains a permanent feature for a Network of Science Centres in South Africa. The currently running training programmes will be continued with until evaluation proves otherwise.

6. Models of the Rollout Plan

Two broad development models, namely the redress model and the public-private partnership model, are advocated for in the National Rollout Plan. The former outlines Government's support for the development of "limited service science centre" while the latter describes the role of private sector in the development of "full service science centres".

The Redress Model

The redress model advocates for the use of limited service science centres as vehicles to achieve the goals of the Network of Science Centres in South Africa. According to National Norms and Standards, a limited service science centre accommodates between 25 000 and 50 000 visitors per annum through direct visits to the centre and outreach initiatives. Such a centre will have a floor space of between 600 m^2 to 1800 m^2. Mobile science centre should be able to reach 50 000 students a year, focusing on rural and remote communities.

The redress model focuses on improving accessibility to science centres by especially under-serviced communities. It entails development of smaller centres which in terms of the National Norms and Standards (2005) has the following elements:

- Phase I of its roll out will entail establish at least one science centre per Municipal District.
- Collaborations with corporate sector will be maximised. This will target organisations' Corporate Social Investment (CSI) funding for various components of the roll out project.

1 Learnership is a tri-cameral arrangement between Government, employer and learner to improve skills and create employment opportunities.

- A science centre in this category will have the following minimum features:
  - *Exhibit hall* to house interactive science exhibits of different science disciplines.
  - *Auditorium* to host science shows, talks, workshops, theatres, etc.
  - *Computer Laboratory* which among others will provide computer literacy programmes to community and create an innovative mathematics and science learning environment.
  - *Career centre* to expose educators, learners and parents to science-based career options; kinds of support available and how to access them (*e.g.* bursaries); potential employers; and on-line career search and applications for admission.
  - *Innovation and Knowledge Development Space* to provide learners with an environment enabling them to apply their basic scientific knowledge to the development of projects relevant to practical problems in their own communities.
  - *Mini-Workshop* to enable a science centre to carry out normal maintenance and repairs of interactive exhibits, as well as research and development of new exhibits.
  - *Mobile Unit* to deliver science centre's outreach activities.

The Public Private Partnership (PPP) Model

The PPP model advocates for the use of full service science centre as a vehicle to achieve the goals of a Network of Science Centres. According to the approved Norms and Standards, a full service science centres has a floor space of at least 1 800 m^2 should be able to cater for 50 000 school visitors per year.

A PPP is defined as a contract between a public sector institution and a private party, in which the private party assumes substantial financial, technical and operational risks in the design, financing, building and operation of a project (National Treasury PPP Manual, 2004). This development model is intended to:

- Afford South Africa an opportunity to have large science centres of international stature, for example, Ontario Science Centre in Canada; and
- To treat South Africans in an equitable and fair manner. The Department is being inundated by proposals for partnerships from individuals and organizations intending to develop large science centres in different provinces. The PPP approach of Government provides criteria for selecting best partners to support the establishment of a Network of Science Centres in South Africa.

Science centres developed under this PPP model will have the following operating characteristics:

- The ideal will be to have one "full service science centre" developed under this model per Province.

- Full service science centres will be developed and operated in terms of Treasury Regulation 16 to the Public Finance Management Act (PFMA), 1999.
- The Department will only describe the output specifications of such a science centre.

7. Monitoring and Evaluation

The Department of Science and Technology appointed the Human Sciences Research Council (HSRC) to manage the tracking of the implementation of the Youth into Science Strategy. This covers the tracking of interventions implemented by the envisaged Network of Science Centres. It is envisaged that a more comprehensive monitoring and evaluation framework will be drafted following recommendations from the tracking study by the HSRC.

8. Conclusions

This paper outlined processes that were undertaken to develop and approve a national policy for the development of a Network of Science Centre in South Africa. It also outlined the goals that this envisaged Network of Science Centres must achieve. Key among these goals include promotion of science and technology literacy; support for increased participation and performance in mathematics and science; and development of skills capacity to innovate.

While political support for the establishment of a Network of Science Centres is beyond question, the challenge is to translate it into financial and community support that will see the goals of the Network realised in the given time-frames. However, the centrality of Government's role has not only created a conducive environment for the development of science centres for a developing economy such as South Africa, but has also served as an example to other developed countries where the focus of science centre role is slowly shifting more towards education than entertainment. It is hoped that this experience will also spur Governments in other developing countries to play more active roles in the establishment of Networks of Science Centres.

References

Department of Science and Technology,1996. White Paper on Science and Technology. Pretoria South Africa

Department of Science and Technology,2002. National Research and Development Strategy. Pretoria. South Africa

Department of Science and Technology, 2005. National Norms and Standards for a Network of Science Centres. Pretoria. South Africa

Department of Science and Technology, 2007. National Rollout Plan for a Network of Science Centres. Pretoria. South Africa.

DF, 2004. National Treasury Public Private Partnership Manual. Pretoria. South Africa.

Chapter 24

Scientific Research Centres in Iran: An Overview

Akram Ghadimi[1] and Tahmineh Shaverdi[2]

[1]*Institute of National Research for Science Policy, Iran*
E-mail: akramghadimi@yahoo.com
[2]*Iranian Academic Centre for Education, Cultural and Research (ACECR), Iran*
E-mail: tshaverdi@msn.com

ABSTRACT

Research and technology are the main factors in achieving the objective of knowledge based development and making it operational. This is one of the main goals in the Islamic Republic of Iran's 20-Year Vision Document as well as in the Fourth 5-Year Development Plan. Considering the low number of researchers as compared to the total population of the country, attention was paid to this important issue in the Vision Document and Development Plan. Thus priority was given to the quantitative and qualitative growth of scientific-research centres. The research and technology sector of the country covers a vast spectrum of academic units and research centres, affiliated to the Ministries of Science, Research and Technology and Health, Treatment and Medical Education. It also includes institutions affiliated to other ministries and executive bodies, research and development departments affiliated to industries, independent private research units as well as national research institutes, networks, growth centres, science and technology parks, centres of excellence, scientific associations and think tanks. These undertake an extensive range of activities from fundamental to semi-industrial research, reverse engineering, development or know-how transfer. The duties of this section include: 1) Training up-to-date manpower in the process of research and technology, 2) Developing the frontiers of knowledge, 3) Exchanging the idea into new products and procedures, 4) Developing and producing technical knowledge and performing semi-industrial research, 5) Transferring and taking in technology, 6) Conducting studies on increasing the competitive capacity of the country's production and service sectors, 7) Performing applied research for solving the problems faced by the country. This paper tries to offer an overview of the situation of the scientific research centres in Iran by defining the duties of this sector in the country and introducing two research organizations. Therefore, the first part of the paper deals with the goals and duties of the sector and the second part introduces two organizations including National Research Institute for Science Policy and Academic Centre for Education, Culture and Research.

Keywords: *Iran, Science, Research, Technology, Development, Research Centre, Science centres, Scientific temper.*

Introduction

The role and situation of scientific sector is a core and key to the realization of the 20-Year Vision's goals and this requires to establish necessary organizations to improve and expand the borders of science, educate and provide skilled manpower, to prepare grounds for the expansion of software movement, to facilitate research and technology, to guide, direct and supply financial sources, to produce, transfer, absorb, and localize and improve technology. It also needs to build a system for making research achievements fixed into business, to promote entrepreneurship, to increase productivity of the Iranian specialists inside and outside the country to produce science and national technology, to establish, develop and utilize new technologies, to set policies, to manage and supervise in producing national science and technology, to establish an information society, to improve and develop recognition of the country and global community and to remove the barriers in the path of development (Document of Scientific Development, 2007).

Cases of desirable perspective with the ideal goals of the country in the 20-year horizon in which, the research and technology play determining role for realization can be summarized as follows:

- A society benefiting from advanced knowledge, is capable of producing knowledge and technology, relying on superior share in human sources and social capital in national production (Article 2 of the Vision Document).
- To achieve first rank in economy, science and technology in southwest Asia with emphasis on the software movement and producing science, accelerated and continuous economic growth, relative improvement of per capita income and realizing full employment (Article 6 of the Vision Document).(Iran's 20 Year Vision Plan, Tehran, 2003)

There are many research units and centres developed in Iran to attain the aforementioned objectives. First, a brief definition is offered on these research centres.

Definitions

The Research Academic Units are defined as the organizational research structures affiliated to the university and conduct research in a specific field in a specialized form. The range of research in these structures consists of followings:

- Basic, applied and development research projects with focus on basic and applied research;

- Research activities for industries and executive apparatus in the form of consultation and conclusion of contract research.

Independent Governmental and Private Research Units

These are specialized and professional units that conduct research in a specific area in a specialized form. The range of activities of these units is mostly focused on applied and development research and as per they might also work in semi-industrial research. (The governmental and public non-governmental research units usually conduct their basic research through joint ventures with the universities or assigning projects to the latter).

Research and Technology Staff Units in Executive Apparatus

These are units that work under the title of deputy office, management, office, etc. in the executive bodies and their duty is to operate research and technology, including policy making, planning, supervision and assessment. Identifying the short term, mid-term and long-term research needs of concerned executive bodies in order to assign project to them and concluding contracts with the qualified research units, receiving the results, cooperation, collaboration and using the results in relevant executive bodies are in the scope of activities of these units.

The R&D Units Affiliated to the State-Run Industries and Companies

The research and technology units with a mission approach and interact in specific specialized field through organized relationship and defined relations.

Research Units Affiliated to the Industries, Production Units and Executive Bodies

These are units that in organizational terms are affiliated to industries, production units and executive bodies and after obtaining permits of the council for improvement of higher education, conduct research in a specialized form in a certain area. Performing applied and development research, reverse engineering and in cases, semi-industrial research to respond to the short-term, mid-term and long-term needs of relevant bodies or organization are among the activities of these units.

The Private Technology Development and Engineering Services Companies

These are companies that mainly aim at the establishment, transfer and localizing technology. The range of activities of these companies is to perform applicable, developmental, semi-industrial, design and reserve engineering research with more emphasize on design, reverse engineering and semi-industrial research.

Growth Centres

These are centres that are formed under the administration of professional specialists and provide supporting services including establishment and development of new professions by entrepreneurs, created in form of newly developed active units in various areas leading to technology and follow economic goals based on knowledge and technology. These centres are not allowed to employ researchers and perform direct research and their duties are mainly support and grow the units established as (nucleus and newly developed companies) in the relevant research activities.

Science and Technology Parks

These are organizations that are run by professional specialists and their main goal is to increase wealth in the society through encouraging and promoting the culture of innovation and improve competition potentials among companies and institutes that rely on science and knowledge at work in the park environment. In order to create motif and manage knowledge and technology flow among universities, R&D centres, private companies and market, the science and technology parks facilitate the process of generating companies and growth centres. They also provide services with high added value, work environments and suitable and quality institution to the established bodies and institutes. These parks are not allowed to employ researchers and perform direct research and their duties are not of the same nature as research institutes and instead, they shall provide overall support and back up of the small and mid-sized companies established in the park.

National Research Institutes

These research units are established by the permit of the Ministry of Science, Research and Technology for national missions. They perform research activities in certain areas and specific goal with national priority. These institutes are allowed to obtain permits and establish necessary units to perform research missions included in clause B of Art 46 of Fourth Development Plan in the relevant grounds. These institutes are also permitted to obtain necessary permits and establish regional units or collaborate with the existing research units in national level to fulfill the duty of the institute in national dimensions.

Networks

These are series of research and technological units with parallel missions and coordinate, interact and collaborate in a specific specialized field in organized and defined forms. The networks are not allowed to employ researcher and perform research directly. (Plan of Development "Research and Technology", 2006)

Centres of Excellence

Scientific pole is an academic (educational or research) centre with scientific activities and significant specifications in educational, research and academic-international collaborations and fields. (<http://www.ut.ac.ir/fa/menues/about excelence.htm>)

Museums of Science and Technology

Museum of science and technology is a place that collects and protects the various works, samples and information related to the scientific, cultural and technological activities and achievements, as the representative of the society and exhibit to public in form of fair collections along with necessary explanations. By employing different methods of informatics and scientific data archives, they enable the visitors to benefit from the information according to their need, their interests and scientific/professional experiences. In addition, the museum holds various scientific and educational programs and improves the area for establishing and propagating scientific thoughts in the society. Museum has a legal entity affiliated to the Ministry of Science, Research and Technology with scientific, non-trading and non-beneficiary nature. (http://www.instm.ir)

Think Tanks

A Think Tank is a non-governmental, non-beneficiary or beneficiary unit that acts totally independent in offering views and perspectives and its goal is to study, make ideas and conduct research for making decisions and setting policies in cultural, social, political, economic and defense grounds and to provide ideas and solutions for the problems and challenges, to provide views on future perspectives for managers and policy making applicants (governmental or private sectors) and contribute to decision making for them. Think tank is one of the research and development units that usually conduct study or research on order. (Plan of Development "Research and Technology", 2004)

Scientific Associations

These organizations are developed and based on agreement and volunteer participation of a group of specialists in a scientific discipline. The scientific associations are not affiliated to the government and are independent in organizational terms and are run based on their article of association and by their own elective managers. (Report Activities Scientific Association, 2002)

Research Status in Development Plans

Research is activity that ultimately leads to production and generation of science and or establishment and completion of technology and in case, it strengthens and reinforces economic-social and cultural fundamentals of the country. Therefore, the research sector plan or in another word, science and technology make one of the main foundations of development plan in any country.

Determining research system of the country is one of the most important plans of research sector. The research plan in Iran was prepared for the first time in 1989 and was approved by the budget and planning commission of Islamic Legislative Assembly. This plan paved way for more attention to research, science and technology in future; for, first, a specific place was allocated to research in the national plans and it became another economic, social and cultural activities unimportance. Second, a part of national resources were deducted to this plan as program. (Yahyaei, 1994)

Supporting open and free research and reinforcing and establishing non-governmental research centres have been among basic policies of research section in the first development plan. Publicizing the research and efforts in determining suitable place for research in society is among other polices of this sector. (Yahyaei, 1994)

In the second development plan special attention was also paid to research as exemplified by the following:

- Expansion and growth of scientific and technical potentials of the country to achieve self-reliance
- To attend the fundamental research as substructure of applied research and improvement of technology
- Expansion of research in strategic fields
- Conformity of research plans and basic needs of the country
- Improve in quality and quantity indexes of research
- Propagation of using the results of research through education and propagation (The Second Plan of Development of Society and Economy, 1995-1998)

In the third development plan, priority was given to research whose most important cases include:

- To establish staff organization and suitable policy making for improving research and technology
- To establish areas of participation and investment of non-governmental sector in research and technology activities particularly research and technology for development application and supporting quality and quantity expansion of research activities of this sector
- To increase demand approach research in governmental sector
- To develop strategic polices and priorities in technology range through determining the trustee and administration of technology section in the country, as ministry of science research and technology as its director, and to change the duties of this ministry with respect to this fact.
- To establish areas of participation and investment of non-governmental sector
- To develop legal bill of act on spiritual ownership right to maintain physical and intellectual rights of innovators and explorers. (Third Plan of Development of Society, Economy and Cultural the Islamic Republic of Iran, 1999)

The Qualitative Goals of Research in the Fourth Development Plan

- To establish national and strategic policy making institute for research and technology
- To develop and increase the potential of research and technology units and improve academic status of at the country
- To facilitate research and innovation and supply financial and human sources
- To establish commercializing system for research achievements, propagate technology and achieving effective role of non-governmental sector
- To achieve desirable level of international interactions in science and in transfer, absorption and localization of technology

- To establish necessary capacity to develop advanced technologies (particularly information, Bio- and Nanotechnologies) and improving entrepreneurship.
- Quality goals: Development of borders of science and improving abilities of universities and research institutes. (Mirbafghi and Nayeri, 2005-2009)

Grand Policies (20-Year) for Scientific Development of the Country

In order to enhance the role of universities and research centres in promoting the productivity of the country, the quality goals of scientific development in the five-year horizon have been defined as follows:

- To develop borders of science and improve potentials of universities and research institutes
- To develop research systems as per the preparation of land and making universities and research institutes mission/assignment based
- To improve the efficiency and effectiveness of scientific and research centres of the country
- To develop research units and improve scientists in the country
- To develop human and social science research, Islamic studies, arts and inter-disciplinary and out-disciplinary majors
- To develop comprehensive system for the supervision, assessment and validation of research in the country
- To establish active relationship in scientific areas across the world, region and Islamic countries. (Plan of Scientific Development, 2007)

Table 24.1: Situation of Research Centres in Iran

Row	*Type of Unit*	*Number*
1	Academic research	143
2	Independent governmental research	128
3	Number of Governmental Research Units	262
4	Private independent research	118
5	Research/dependent on other governmental organizations	66
6	Centres of science and technology growth	37
7	Science and technology parks	15
8	Think Tanks	3
9	Centres of excellence	69
10	Networks	4
11	Museum of Science and Technology	23
12	Number of scientific associations	175
13	R and D	444

The National Research Institute for Science Policy

The National Research Institute for science Policy (NRISP) was founded in 1980 under the auspices of the Ministry of Science, Research and Technology (formerly the Ministry of Sciences and Higher Education).The appointments were based on the NRISP Constitution approved by the Ministry.

Goals of the Institute

The general goal of the National Research Institute for Science Policy is to study in scientific, academic and research policy making, research and study for identifying needs and possibilities of national research system and performing studies needed by the Ministry of Science, Research and Technology for scientific and research policy making and development programs of the country.

Functions and Powers

1. To study and identify research needs in scientific policy making and developing scientific activities of the country;
2. To carry out fundamental, applied and development research plans to realize the goals of the centre;
3. To study on the policy making and planning for the culture of science propagation in the society
4. To provide scientific, technical and managerial consultation services in scientific policy making;
5. To provide necessary facilities that would fit the relevant research activities;
6. To establish constructive interactions with specialized forces and scientific and research institutes inside and outside the country to realize the goals of the centre within the framework of regulations and rules.
7. To propagate and publish results of studies and research performed, in form of scientific journal or books, electronic sits and similar means according to the regulations of the country;
8. To present research achievements in form of short term educational workshops without awarding official academic degrees;
9. To plan and hold scientific conferences in the relevant fields in local, regional and international levels according to the criteria and regulations;
10. To plan and hold scientific conferences science propagation festivals in the society
11. To perform administrative, financial and transactional activities related to the duties of the centre according to the regulations and criteria of the country.

Description of Functions of the Institute

- To determine main lines of scientific policy of the country with respect the social, cultural and economic goals and economic development plans and presenting them to concerned officials;

- To study on scientific regulations and rules in connection with research institutes and activities
- To research in scientific and research planning and comparative studies in other countries
- To study the quality of output increase in research system
- To study on scientific research stats in international level
- To study for knowledge of scientific potentials of the country and university and non-university institutes and centres
- Study in the area of cooperation and research exchanges with other countries
- Scientometric studies for constant evaluation of scientific activities and determining scientific status of Iran in the world
- Studies on research credits in various aspects and suitable approach for better use of research credits
- Study on the method of improving science in the society
- To train specialized human forces in science policy making
- To study and research in foresight of science in the country
- To perform necessary studies and actions in supporting, providing grounds and expanding research attitude in the country.

Fundamentals

A. Board of Trustees

B. President of Research Council

C. Research Council

Research Council

President of the centre (chairman of council)

Research deputy of the centre (secretariat of the council)

Managers of research groups

Two to five faculty members

Function and Powers of the Research Council

The function and powers of the research council include approval of internal by-law of the council, to determine policies and scientific and research programs of the centre in the framework of policies of the ministry, to approve by-laws and research instructions of research groups of the centre, and to study and approve research plans of the centre and supervise the quality and their good performance. Moreover, to study and supervise the progress of research plans of the centre, to determine arbitrator or arbitration council for the research plans completed at the centre, as introduced by the research deputy, to evaluate the results of scientific and research activities of the centre, to approve suggestion for the establishment and dissolution of research groups of the centre in order to meet legal formalities, to plan for establishment and deepening scientific relationship with other institutes and educational and research centres of the country according to the relevant regulations and laws are among the other powers of the research council. Other functions of the council can be summarized as following: to study and approve short term educational courses, conferences festivals and scientific visits, to study and approve scientific trips, educational scholarships and sabbatical leaves for faculty members in accordance with regulations and laws, to study and determine by-laws for writing and translating books, papers and method of publication results of scientific and research activities, to study and determine facilities, publication, informative and scientific needs of the centre and suggest their supply by head of the centre, to study the scientific qualification of applicants for membership at the faculty member of the centre and expressing views on it, and finally to perform other affairs which are assigned to the council or head of the centre by the board of trustee in the framework of criteria and regulations.

Initially, NRISP consisted of six scientific groups, namely the Foresight group, the Popularization of Science group, the Science and Society group, the Science Policy group, the Economy of Science group and Scientometric group. These groups were headed by professors from respective fields.

Foresight Group

To study the status of science and technology in Iran and drawing its perspective in future

To determine the research priorities in connection with future needs.

Popularization of Science Group

- Studies in propagating science in the society:
- Planning and policy making on the quality of science propagation in society, particularly in youth level
- Publication of books, journals and gazettes for propagating science in simple language

Science and Society Group

- To study on the role and importance of scientific and technological activities in the progress of society and development programs
- To study the best methods and ways of optimized use of management of research institutes and management of human resources in research centre
- Science sociology studies
- To study different issues and barriers in the path of research expansion.

Science Policy Group

Economy of Science Group

- The studies on the counter role of economics and research
- Study and research in making the results of research commercial

- To define the indexes related to research credits and measuring them in national and international scales
- Define indexes on scientific and technical personnel

Scientometric Group

- Quantitative study of science and technology based on international databases
- To define indexes needed to measure scientific potentials, quality and quantity changes of science and conscious structure science in different national, regional and international levels
- Study in scientific and research classification and higher education centres and research institutes
- To prepare national reports on research. (http://www.Nrisp.ac.ir)

Academic Centre for Education, Culture and Research (ACECR)

Following the command of the leader of the Islamic Revolution, Imam Khomeini on June 13, 1980 concerning establishment of cultural revolution headquarter, the mentioned quarter was established in the session dated August 7, 1980 to realize the goals of cultural revolution and established ACECR as a revolutionary institute generated from cultural revolution. After conversion of the staff to high council of Cultural Revolution, the article of association of this body was revised on December 2, 1986 and was approved by deleting the duty of participation in management of universities and emphasizing on cultural and research activities to Islamize universities. Once again, Cultural Revolution high council, with respect to the new conditions of universities approved a resolution on November 13, 1990 through which, ACECR was introduced as a bridge between universities and industrial and service sectors. In this resolution, the two main duties of scientific studies and cultural activities were remained under control of this body; Based on this resolution, the new article of association of ACECR was approved by high council of Cultural Revolution as follows:

ACECR is a public and non-governmental body under superintendence of high council of Cultural Revolution and has independent entity and in terms of administrative, employment and financial issues, it follows regulations which are approved by the board of trustees of ACECR within the legal framework.

The Goals of ACECR

- To expand research and flourish the attitude of search and find talents in the society level to achieve self-sufficiency
- To develop cultural affairs and issues in society through cooperation with the theology school of Qom, university and other cultural bodies and institutes particularly for young generation
- To expand applied and semi-industrial plans through links with scientific and research centre to use the results of the research.

The General Duties of ACECR

1. To propagate and advertise Islamic culture and arts and make cultural and educational ideological planning and establishing mental and ideological preparation in young groups especially students of high schools and colleges to confront and fight western thought and deviations in the cultural policies of the country.
2. To build suitable establishments for writing, translating and publishing books and journals and releasing the results of investigators and scientific, cultural articles and producing educational materials
3. To perform developmental and applied research
4. To follow up research plans of jihad up to the semi and full industrial stages
5. To present scientific- technical services in different areas needed by the society
6. To support, encourage and attract students and young and talented researchers based on knowledge of actual problems and specialized needs of society
7. To execute official educational courses in scientific-applied fields in accordance with the regulations and laws of higher education of the country
8. To establish educational and research organizations as mentioned in the clauses of this article based on higher education regulations and laws
9. To participate in establishing suitable grounds for more occupation of university graduates

The areas of this participation cover scientific and technical services including performing quality tests, studying the suggestions of standards, collection and processing scientific data, publishing scientific data and documents, propagating techniques and performing scientific construction, presenting technical contributors and performing health and medical services and medical diagnosis. (http://www.Acecr.net)

Conclusion

This research describes the existing conditions of scientific-research centres in Iran. The aforementioned research centres are established and work within the framework of the Vision Document and development plans; hence efforts are made to achieve the goals as noted in the aforementioned documents. Different research centres in Iran work in line with scientific development and their important and critical role in Iran's scientific development is undeniable.

References

Document of Scientific Development, 2007. Ministry of Science, Research and Technology, National Research Institute for Science Policy and Department of Research Ministry of Science, Research and Technology, pp. 46.

<http://www.ut.ac.ir/fa/menues/about excelence.htm>.

http://www.instm.ir

http://www.Nrisp.ac.ir

http://www. Acecr.net

Iran's 20 Year Vision Plan, Tehran, 2003.

Mirbafghi, Seyed Hossein and Nayeri, Ebrahim Moradiyan, 2005-2009. Law of Fourth Plan Development Economy, Society and Culture, The Islamic Republic Iran.

Plan of Development "Research and Technology", 2006. Organization of Management and Planning and Ministry of Science, Research and Technology, Teheran, pp. 8-9.

Plan of Development "Research and Technology", 2004. Ministry of Science Research and Technology, pp. 10.

Plan of Scientific Development, 2007. National Research Institute for Science Policy, and Deputy of Research, Ministry of Science, Research and Technology, pp. 53.

Report Activities Scientific Association, 2002. Commission of Scientific Association and National Research Institute for Science Policy, Tehran, pp. 3.

The Second Plan of Development of Society and Economy (1995-1998). Part of Research, 3 Index.

Third Plan of Development of Society, Economy and Cultural, 1999. The Islamic Republic of Iran (1999-2004). Organization of Management and Planning, Tehran.

Yahyaei, Fariba Fahim, 1994. Research Indexes and Plan First and Second, Rahyaft Journal, No 7, Tehran, pp. 18.

Chapter 25

Sustainable Research through International Cooperation

Tran Dung Tien*, and Luc Gia Thai**

Sweden Research Cooperation Programme, Ministry of Science and Technology, Hanoi, Vietnam

*E-mail: * tdtien@most.gov.vn, ** thailucgia@yahoo.com*

ABSTRACT

The science and technology centres in Vietnam have changed considerably in terms of functions, organization, funding mechanism, incentives for researchers, etc. over the past few decades. The long-term plan for Vietnam envisages that it should become an industrialised, modern country by 2015 through the development of science and technology and application of its results. With this purpose the bilateral Programme of Research Cooperation between Vietnam and Sweden started in 1977. The main objectives of this Programme are: to strengthen the research capacity, to provide financial and scientific resources to produce new knowledge by conducting joint research on relevant topics, to develop viable networks of international research cooperation. This paper focuses on the "sandwich model", which, within the programme, has been an effective way to develop and strengthen university and institutional cross-border cooperation. The so-called "sandwich model" has been successfully used for research training and as a base for other research cooperation.

Keywords: *Vietnam, Science and Technology, Research, International Cooperation, Sandwich model, Capacity building.*

Role of Science Centres for Development in Vietnam

The long-term plan for Vietnam envisages that it should become an industrialised, modern country by 2015 through the development of science and technology and application of its results.

For accomplishing this task there is a need to enhance the role of science and technology centres which include: (1) change in the institutional framework, (2) reform of science and technology management, (3) change in the funding mechanisms, (4) focus on human resource development, and (5) enhanced international cooperation.

The science and technology centres in Vietnam have changed considerably in terms of functions, organization, funding mechanism, incentives for researchers, etc. over the past few decades. It is expected that this change will continue depending on the market economy. There are several issues relating to cooperation between universities and research institutes, the control of research priorities, and cross-sectoral co-ordination of priorities that should be discussed, evaluated and reformed anew.

The Vietnam–Sweden Research Cooperation Programme

The bilateral Programme of Research Cooperation between Vietnam and Sweden (hereinafter referred to as programme) started in 1977. Over the last 30 years, approximately SEK 280 million granted by the Swedish government have been used to finance research cooperation of about 30 Vietnamese and Swedish universities and research institutes. In the initial years, the programme focused on international exchanges, but since the late 1980s, the focus has been put on research capacity building. Many topics have been included, but health research, agriculture and rural development studies for pro-poor programmes have been prominent all the time.

The main objectives of this Programme are:

- To assist Vietnam strengthen research capacity in the form of creating conducive research environments and providing research education,
- To provide financial and scientific resources to produce new knowledge by conducting joint research on topics that are of importance to both Vietnam and Sweden and transfer such research results to the users in the development process, and
- To develop viable networks of international research cooperation with a view to assisting Vietnamese scientists and researchers to integrate proactively into the international research community.

The Programme has had a significant impact on the Vietnamese society. Impact has mainly been on the economic and environmental aspects. The researches on antibiotic resistance, Shigella vaccine, drug resistance of malaria, common diseases in Vietnam, etc., have obtained significant results, bringing Vietnam self-sufficiency in some types of vaccines so that the vaccines may even be exported.

Forestry Research developed hybrid clones of Acacia and Eucalyptus, as well as pine species which are widely used in the government reforestation programme, as they are three to four times as effective (grow faster and produce better timber). The new hybrids have brought significant results to the forestry sector. These researchers have been recognised nationally and awarded medals for their contributions.

Farming systems research has developed new knowledge of animal feed and husbandry practices, as well as new types of biogas converters. These are applied widely in provinces of the Mekong delta in the South, and also transferred to the countries in the region.

Research on marine environment has not only supported decision-makers in integrated coastal management planning but also farmers in finding ways of cultivating molluscs, shrimps without causing damage to the environment.

Other projects have also had an impact on government policy formulation and legislation (Health systems research and Science and Technology Policies).

With regard to research education within the Programme, by the end of 2006, 28 PhD and 33 Master students had successfully defended their dissertations under the "Sandwich" training model, explained below. It is expected that by the end of 2010, 46 PhD and 25 MSc students more will have completed their studies under this mode. Besides, hundreds of shorter training courses have been conducted for newly-graduated students and junior researchers at Vietnamese universities, institutions. Many researchers have had opportunities to attend international and regional scientific conferences, workshops and seminars.

"Sandwich Model" in Research Training/Education

This paper focuses on the "sandwich model", which, within the programme, has been an effective way to develop and strengthen university and institutional cross-border cooperation. The so-called "sandwich model" has been successfully used for the research training and as a base for other research cooperation. In this model the research training student carries out her/his research project about half of time in their home department and the other half in the department in Sweden and the student has supervisors from both sides.

The studies thus would be performed for some periods in Sweden to explore advanced resources over there and for some periods at the home universities to associate with the country needs. Furthermore the cooperation will be stimulated and supported by supervisors and departments of both sides thus enhancing cooperation between universities/research institutions. Consequently, the student never leaves the home department and will be able to continue to be active at the home university. The idea is that the student remains within her/his own departmental environment and therefore will more likely formulate research frameworks that are relevant and sustainable to the home country.

The research training programme is planned to last for 4-5 years for PhD and 2-3 years for MSc. degree at the Swedish collaborating universities and in Vietnam. Necessary preconditions for a smooth functioning model are active coordination from each university and continuous interaction between the supervisors at each side. The program support therefore includes opportunities for the students' supervisors to visit the other university once a year in order to build up a network among the partner universities. The other part of the cooperation is aimed at establishing long-term relationships at all levels of the university, up to the presidential level.

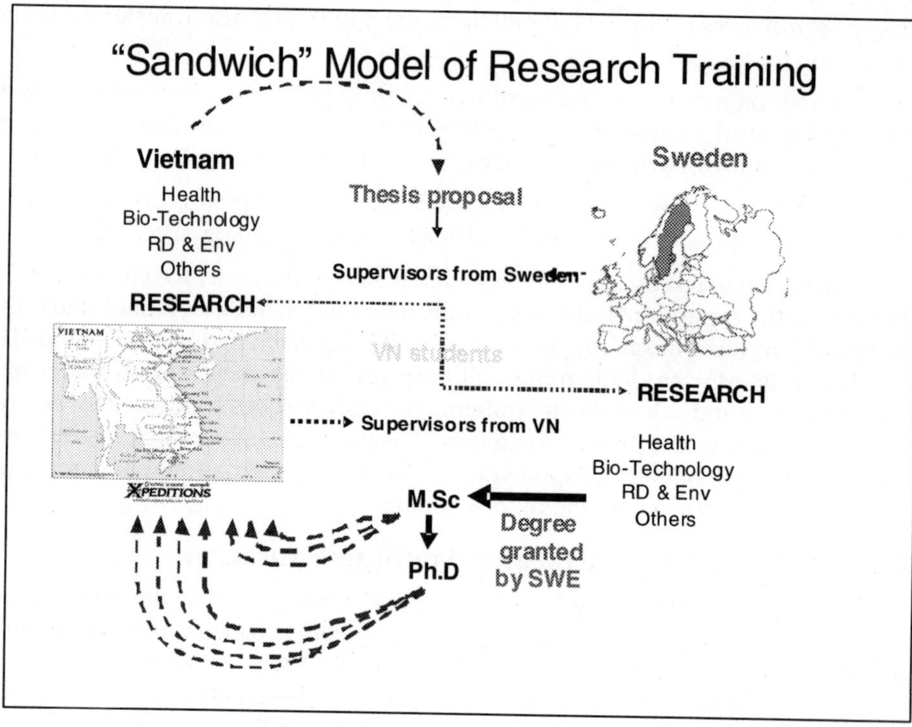

Figure 25.1: 'Sandwich' Model of Research Training

Benefits and Advantages

Students graduated from the Programme are now generally in good position, having returned to their home universities as lecturers, working as research officers, research coordinators or head of research stations, deputy director of a school of agriculture, whereas others have found positions in ministries or are working in the extension service sector.

The PhDs and MScs trained within the Program and graduated from Swedish universities (SLU, KI etc.) have also been successful in their employment. They have returned to higher positions after their training as senior researchers, key lecturers, heads of research centres, deans of faculties, universities and also having shown an ability to manage advanced and complex tasks. Most of them are now considered important resource persons in their respective departments/faculties/organizations.

Conclusion

The research capacity building is a complex and long term undertaking. The program has contributed and is contributing in terms of both individual training and institutional improvement and other items such as equipment needed for research, good team-work spirit, more advanced teaching and research methodology. There is evidently still a considerable lack of capacity in some of the universities and research institutes in the network, it is therefore necessary that the capacity building process that the programme has developed should continue. Effectiveness of capacity building is closely related to institutional strengthening. Without necessary organizational changes, skills that have been acquired may not be used and would soon be eroded.

Chapter 26

Achievements of a Research Council: The Case of Zimbabwe

Susan Muzite
Research Council of Zimbabwe, Harare, Zimbabwe
E-mail: secretary@rcz.ac.zw, susan@rcz.ac.zw

ABSTRACT

The achievements of Research Council of Zimbabwe (RCZ) from 1984 to 2007 are presented taking into account the international, African and regional, and national contexts. The RCZ was established in terms of the Research Act Chapter 10:22 to promote, direct, supervise and coordinate research. To that end the RCZ has given advice to Government on the Science and Technology Policy, set up institutions, held eight symposia, given grants and awards and registered foreign researchers. The future of RCZ rests with management of the national research agenda, the nature of which is still being worked out in the light of the Ministry of Science and Technology (MSTD) new mandate.

Keywords: *Research Council of Zimbabwe, Science, technology and innovation, Research and development, Research promotion, Foreign research registration, National research agenda.*

Zimbawe Science and Technology

Research Council of Zimbabwe (RCZ)

RCZ Mandate

The RCZ was established in 1984 to promote, direct, supervise and coordinate research. One of the major functions of RCZ is advising Government on issues of research for sustainable development. RCZ also provides an exceptional forum for interaction and discussion for the mutual benefit for Government, academia and industrialists. In addition, it is a convenient conduit for financial and infrastructural support for collaborative research among research institutions and councils. RCZ's role is that of a catalyst, having identified broad areas of concern, it consults and brings together relevant experts to define a programme of work and to seek out sources of funds. The RCZ establishes and maintains links with professional bodies and centres of excellence to enhance the quality of its role as facilitator of national and international collaboration.

Board Membership

In terms of section 4 of the Research Act, Chapter 10:22, Council members are appointed by the Minister and shall not be less than ten (10) and not more than fifteen (15) in number and one such member is appointed as Chairman and one other member as Vice-Chairman. Membership is for three years.

Committees

The Standing and other committees are the operational arms of the RCZ and are chaired by members of Council. RCZ Standing Committees make sectoral considerations and technical inputs of issues before Council and may liaise with relevant local and international specialist organizations. Committees are also responsible for the implementation or follow-up of Council decisions. Implementation occurs when the committee members collectively or individually have to implement a Council decision. Follow-up occurs when the Secretariat or other agencies are implementing and the committee is responsible for ensuring that objectives are met.

The Chairperson of every Committee is accountable to Council on all matters discussed, implemented, facilitated or undertaken in any form by the committee. As such, Chairpersons periodically report to Council on the activities of their committees. Other members of the Research Council Board may, on their own volition, attend meetings of any standing or other committee.

There are nine (9) Standing Committees and four (4) other Council committees which cover the following sectors;

Standing Committees

1. Science and Society
2. Agricultural Sciences
3. Natural and Environmental Sciences
4. Industrial Development
5. Mineral Resources and Earth Sciences
6. Health Sciences
7. Social Sciences and Humanities
8. Informatics
9. Foreign Researchers

Ad Hoc Committees

1. Finance and Administration

2. Research Grants and Awards
3. Biennial Symposia (currently 9th Symposium)
4. Bioethics

Secretariat

The Secretariat is the executive arm of the Council responsible for the day-to-day operations and is headed by an Executive Director.

Vision

To be the pivotal leader, guiding all research towards the sustainable development of Zimbabwe.

Mission Statement

To provide leadership in research for national development.

Core Values

Quality, Professionalism, Integrity, Research ethics and Innovation

Structure of Paper

The paper seeks to situate the RCZ within the global, continental, regional and national contexts. It is within its contextual environment that the Council has sought to deliver on its mandate and achieve results proactively. The achievements are listed according to the categories of policy advice to Government and promotion of research. The final part of the paper looks at the future of the RCZ in the field of management of the national research agenda (R&D) together with innovation for sustainable socio-economic development.

The Contextual Situation of RCZ

Global and International

At the international level the RCZ attends United Nations Educational Scientific and Cultural Organisation (UNESCO) General Conferences and is a member of the International Council for Sciences (ICSU). Through participation at General Conferences of UNESCO, RCZ has become aware of issues addressed by commissions and programmes such as:-

- Intergovernmental Oceanographic Commission (IOC)
- International Basic Sciences Programme (IBSP)
- International Geoscience Programme (IGCP)
- International Hydrological Programme (IHP)
- Man and the Biosphere Programme (MAB)
- Management of Social Transformations Programme (MOST)

Within that context issues of particular interest to the RCZ are:

- Climate change
- Hazards and disasters

- Human well-being
- Environment sustainability
- Management and sustainability of natural resources

These are in tandem with the UNESCO Biennial Sectoral Priorities for Natural Science namely:

- *Biennial sectoral priority 1*: Promoting research and technical capacity-building for the sound management of natural resources and for disaster preparedness and mitigation
- *Biennial sectoral priority 2*: Strengthening national and regional research and innovation systems, capacity-building, the use of technologies, and scientific networking, and encouraging the development and implementation of science, technology and innovation policies for sustainable development and poverty eradication

The RCZ is a member of ICSU and participated actively in formulating its Strategic Plan covering the period 2006-2011. ICSU identified the following major issues in planning and coordinating research:

Environment

- The international Polar Year (1Py 2007)
- Natural and human induced environmental hazards and disasters

Sustainable Development

- Sustainable energy

Human Health

Exploring New Horizons such as:

- Cognitive neurosciences
- Complex systems science
- Human security
- Environmental refugees,
- Molecular biosciences
- Nanosciences and nanotechnology
- Transgenic crops and animals (ICSU, 2005)

Mention must be made of the RCZ affiliation to the NAM Centre for Science and Technology. Zimbabwe has benefited from training workshops and scientific meetings of the Centre. Zimbabwe young scientists have also received awards in the fields of Health, Mathematics, Physics, and Medicine offered by the Academy of Sciences for the Developing World among many human resources development programmes and initiatives of the Academy.

Continental Level and Regional Level

Africa's Science and Technology Consolidated Plan of Action of August 2005 is premised on three pillars namely capacity building, knowledge production and technological innovation.

The flagship programmes of the Action plan are:

Cluster 1

Biodiversity, Biotechnology and knowledge

Cluster 2

Energy, water and desertification

Cluster 3

Material sciences, manufacturing laser and post–harvest technologies

Cluster 4

Information and communication technologies and space science technologies (African Union, 2005).

The Lagos Plan of Action for the Economic Development of Africa, 1980-2000 also guided the views of RCZ with respect to self-reliance and sustainable development. The ICSU regional office for Africa identified four priority areas in energy, health and human well-being, natural and human induced hazards and disasters, and global change (ICSU Regional Office for Africa, 2006).

RCZ has remained watchful of the initiatives of the Southern African Development Community (SADC) in research, science, technology and innovation.

National Level

Zimbabwe gained independence from colonial rule in 1980. Government sought to meet socio-economic development aspirations through Growth with Equity Strategy of 1981, Zimbabwe Transitional National Development Plan (1982-1985), Zimbabwe First Five Year National Development plans (1986-1990), Economic Structural Adjustment Programme (1991-1995), Zimbabwe Programme for Economic and Social Transformation (ZIMPREST) of 1998, Millennium Economic Recovery Programme of 2001, National Economic Revival Programme of 2003, Macro-economic Policy Framework: Towards sustained economic growth (2005-2006), National Economic Development Priority Programme, and Zimbabwe Economic Development Strategy (ZEDS 2009-2013) in 2007. These stated policies, plans, strategies, frameworks and programmes demonstrated the country's desire and concerted effort with respect to socio-economic development (Ministry of Public Service, Labour and Social Welfare, November 2006). The Research Council of Zimbabwe has been part of that effort and registered achievements.

Achievements of the Research Council of Zimbabwe

Policy Advice to Government

The Research Act Chapter 10:22 was promulgated in 1986 on the advice of RCZ. Over a ten year period, the RCZ conducted an extensive stakeholder consultation process on a science and technology policy for the country. A document was submitted for the consideration of Government in 2001 and was passed by Cabinet in November the same year. His Excellency the President C de R. G. Mugabe launched the Policy at an especially arranged event attended by about 4 000 people from all sectors of the economy in June 2002. The policy covers the following:

1. Research and Development for industrialisation
2. Agriculture, food security and environment conservation
3. Mineral Resources and Earth Sciences
4. Education, Training and Manpower Development
5. Health
6. Energy
7. Transport
8. Information and Communication Technology
9. Housing
10. Biotechnology
11. Role of Social Sciences
12. International Collaboration (Research Council of Zimbabwe, March 2002)

In August 2002 a Department of Science and Technology Development was created to ensure implementation of the policy. The Department became a full Ministry in 2005.

Promotion of Research

Under the key result area of promotion of research the RCZ has established institutions, hosted symposia and given grants and awards.

Institutions

RCZ established the following institutions:

- 2004: Zimbabwe Academy of Science (ZAS)
- 2000: Bio-safety Board (now transformed into the Biotechnology Authority of Zimbabwe)
- 1997: Zimbabwe Academic and Research Network (ZARNET)
- 1993: Scientific and Industrial Research and Development Centre (SIRDC)

Symposia

The objectives of symposia are:

- To provide a platform for the promotion of research for national development.

- To enable researchers and policy makers in Zimbabwe and the world at large to exchange information, research findings and ideas.
- To promote smart partnerships among researchers, industrialists and communities to ensure application of research findings.

Papers presented are published in referred proceedings to the standard of being accepted for promotion at academic institutions.

RCZ has hosted symposia as follows which were attended by regional and national participants:

- 2007: "Knowledge Based Development for Zimbabwe"
- 2004: "Impact of Innovative Science and Technology on National Wealth Creation"
- 1999: "A Century of Science and Technology Challenges for a next Millennium"
- 1996: "Towards Capacity Building in Science and Technology"
- 1994: "Advances in Productive and Sustainable Applied Technology"
- 1992: "Innovation, Self Reliance and Development"
- 1990: No Theme
- 1988: No Theme

Research Grants to Young Scientist

These are awarded annually on a competitive basis. The call for proposals is advertised in national newspapers and on the RCZ website. Grants are given in the areas which have immediate impact on national development aspirations.

Grants for Publication of Journals

RCZ supports publication of the following journals:

1. The Central African Journal of Medicine (CAJM)
2. Journal of Applied Sciences in Southern African (JSSA)
3. Journal of Science and Technology (JS&T)
4. Zimbabwe Science News
5. Zimbabwe Veterinary Journal
6. Zambesia Journal of Humanities

Biennial Awards at RCZ Symposia

The following awards are presented during RCZ Symposia:

1. Presidential Award to an individual for outstanding research whose findings have been applied.
2. Presidential Award to a Company/Institution for its outstanding contribution in employing R&D

3. RCZ Sectoral Awards to individuals or institutions for outstanding research in the following sectors:
   - Agriculture
   - Mineral Resources and Earth Sciences
   - Information and Communication Technology
   - Health Sciences
   - Industrial Development
   - Natural and Environmental Sciences
   - Social Sciences
   - Humanities
4. Various Awards to Young Scientists' Exhibitors are given to young scientists who exhibit at science and technology symposia. Young scientists who are mainly high school pupils have been a constant attraction at science and technology symposia.

Registration of Foreign Research

Council has developed application guidelines (Appendix) and an application form. Foreign researchers from all over the world have registered to conduct research over the years. The Committee of Council responsible for considering applications meets monthly.

Management of the National Research Agenda

The creation of the Ministry of Science and Technology Development (MSTD) makes it necessary to re-examine the Research Act for the purpose of re-aligning responsibilities of the Research Council of Zimbabwe separate from those of the Ministry. The re-examination process also encompasses innovation and innovation systems. Principles of a new Science, Technology and Innovation (STI) Bill were approved by Government. The future role of the Research Council of Zimbabwe in the management of the national research agenda is being worked out within that context. In the interim, the RCZ has works in progress with respect to research prioritisation, research audit and database development, bioethics and biospecimen shipment guidelines.

References

ICSU, 2005. ICSU Strategic Plan 2006-2011, Paris, France, pp. 27-33.

African Union, August 2005. Africa's Science and Technology Consolidated Plan of Action, Addis Ababa, Ethiopia, pp. 5.

ICSU Regional Office for Africa, 2006. Annual Report 2006/07, Pretoria, South Africa, pp 1.

Ministry of Public Service, Labour and Social Welfare, July 2006. Zimbabwe 2003 Poverty Assessment Study Survey Summary Report, Harare, Zimbabwe, pp 3-4.

Research Council of Zimbabwe, March 2002. Republic of Zimbabwe Science and Technology Policy, Harare, Zimbabwe, pp 11-15.

Abbreviations

- ICSU: International Council for Science
- MSTD: Ministry of Science and Technology Development
- NAM S&T CENTRE Centre for Science and Technology of the Non-Aligned and Other Developing Countries
- R&D: Research and Development
- RCZ: Research Council of Zimbabwe
- SADC: Southern Africa Development Community
- SIRDC: Scientific and Industrial Research and Development Centre
- S&T: Science and Technology
- STI BILL: Science, Technology and Innovation Bill
- UNESCO : United Nations Education Scientific and Cultural Organisation
- ZARNET: Zimbabwe Academic and Research Network
- ZAS: Zimbabwe Academy of Sciences

APPENDIX

Approval Procedures for Foreign Researchers in Zimbabwe

Introduction

Foreign researchers or any person wishing to conduct research in Zimbabwe on behalf of a foreign institution, foreign organisation or other foreign person should follow the guidelines below:

Application Guidelines

Foreigners intending to conduct research in Zimbabwe or Zimbabwean nationals acting on behalf of foreigners are required to apply for research permits from the RCZ. Such applications must be made and approved before arrival in Zimbabwe in the case of foreigners. The foreign researcher should seek an institution of affiliation in the area of the intended research in Zimbabwe. That institution of affiliation in Zimbabwe forwards the application to RCZ. It is up to the institution to accept the foreign researcher.

All foreign researchers must clearly state the purpose of their intended research on their application forms failing which the application may not be considered. Applications for research in the medical field should be sent to the Medical Research Council of Zimbabwe (MRCZ), P.O. Box CY 573, Causeway, Harare, Zimbabwe (e-mail: mrcz@mrczimshared.co.zw). Approval by the MRCZ is a prerequisite for RCZ consideration of the application.

Research permits are issued for a continuous period of three years. After expiration date the research project may only continue upon renewal. For renewal, a progress report together with registration fees, should be submitted no latter than one month before the expiration date. All applications for extension of research permits are treated as new applications and must be accompanied by a brief progress report of the completed research.

Application Procedure

Six copies of the completed application form, Form No. RA1 Section 26 A, plus one comprehensive research proposal per project must be submitted to The Executive Director, Research Council of Zimbabwe. For joint projects, one comprehensive research proposal is required but each applicant must submit 6 copies of his/her own completed application form.

The researcher must attach letters of support from Zimbabwean institution of affiliation and other institutions he/she may wish to consult with during the course of the research (6 copies of each). Applicants must provide evidence of adequate funds for the period of research in Zimbabwe to cater for his/her personal and research needs. The applicant must pay, through the institution of affiliation or directly to the RCZ, a non-refundable fee which is currently US$500.00.

The Research Permit is issued out to the institute of affiliation for onward transmission to the researcher. Submission of 6 copies of each document submitted to RCZ except for the detailed proposal is a prerequisite for the application to be considered. The applicant is required to submit a detailed report to RCZ within one year of completing the research.

Further information is obtainable from the:

Executive Director,
Research Council of Zimbabwe
P. O. Box CY 294,
Causeway, Zimbabwe.
Tel: +263-4-369407/8, 338258
Fax: +263-4-369409
Email: technical@rcz.ac.zw

Chapter 27

An Analysis of Malawi's Scientific Output

Mzamose Pumula Gondwe,
College of Medicine, Blantyre, Malawi
E-mail: mgondwe@medcol.mw

ABSTRACT

Publications productivity, the number of scientific articles published is a measure of a country's scientific output. If measured carefully it can be a useful indicator that describes a countries' research activity. Our objective was to analyze trends in scientific output originating from Malawi between1996–2006. The MEDLINE/PubMed database was searched for articles originating from Malawi between1996–2006 by typing Malawi in the author affiliation search field. Review of abstracts was performed to determine health field and origin of first author–Malawian vs. foreign. 506 articles were retrieved of which 489 were on health. 15.5 per cent on TB, 14.5 per cent on HIV and AIDS, 11.2 per cent on infectious disease, 7.2 per cent on TB and HIV, 7.2 per cent on Malaria. 20.9 per cent of the authors were of Malawian origin and Tropical Doctor was the journal that had the most articles originating from Malawi. The number of articles published from Malawi has grown by 106 per cent in the past ten years. Our results suggest there is growth in scientific publishing in Malawi but the main contribution is from foreign researchers residing in Malawi. This article clearly highlights the disparity with only 1 out of every 5 articles being authored by Malawian.

Keywords: Scientific output, Publication, Research, Developing countries.

Introduction

Research has the potential to address many challenges that affect developing countries such as poverty, health, food security, HIV/AIDS, malaria, and education. It is for this reason that developing countries should strengthen their research activities. One measure developed by the Organization for Economic Cooperation and Development is to describe research activity as the amount of money that is spent on research and development (R&D), as a proportion of a countries gross national product (GNP). These figures have formed the basis of scientific league tables which allow countries to measure their relative position (http://www.oecd.org/document/26/0,3343, en_2649_34451_1901082_1_1_1_1, 00.html). African science ministers in 2004 agreed at a New Partnership for Africa's Development (NEPAD) meeting to seek to persuade their economic counterparts to increase each countries spending on R&D to 1.0 per cent of GNP. Most advanced nations spend between 2.0 and 3.0 per cent of GNP on R&D; in Africa, the figure tends to be 0.2 to 0.3 per cent. New measures are now being used to measure research activity such as the number of papers published. Despite using this measure, research shows that there is still a large north to south gap in scientific publications originating from Africa but there is a slow but steady increase (Rahman and Fukui, 2003; Perez-Iratxeta and Andrade, 2002).

Studies have been published on quality and quantity of scientific output in African countries (Adamson, 1992; Badrane and Alaoui-el-Azher, 2003; Chatelin and Arvanitis, 1992) however there have been no studies which examine comprehensively the scientific output of authors in Malawi.

This study aims to analyze scientific productivity in Malawi and discuss qualitative aspects of this growth. We evaluate trends in quantity, discipline, journal, and origin of first author to determine contribution of local Malawian authors.

Methods

We used the electronic Medline database even though it favors North-American and European publications, the Medline/PubMed database is the one of the most widely used international database for the health. In order to characterize an article of Malawian origin, we used "Malawi" as the key term in the author affiliation field. The study time period was defined in the "Limits" tab of the PubMed search interface. We reviewed abstracts of the retrieved articles and articles not pertaining to health were excluded. First author's origin *i.e.* Malawian vs. foreign was determined by inspection of his or her name. For cases in which an authors name was uncertain, a third party was asked to determine. We also quantified the numbers of articles originating from Malawi, in various journals per year.

Results

We retrieved 506 of articles for the study period 1996 to 2006 of which 489 were health related. 4 articles were excluded from the analysis as they were considered non-medical as they were methodology papers. 20.9 per cent of the articles had Malawian first authors. Of the 489 health articles 15.5 per cent (76) were on TB, 14.5 per cent (71) were on HIV and AIDS, 7.2 per cent (35) were on TB and HIV, 11.2 per cent (55) were on infectious diseases, 7.2 per cent (35) were on Malaria, 6.3 per cent (31) were on reproductive health (Table 27.1, Figure 27.1). There has been a 103 per cent increase in the number of articles published in 10 years. Tropical Doctor was the journal that had the most articles originating from Malawi followed by Journal of

Tuberculosis and Lung Disease and Transactions of the Royal Society of Tropical Medical Hygiene.

Table 27.1: Number of Publication by Disease Type and Year of Publication

Totals for All	*Totals No. of Papers*	*TB and HIV*	*TB*	*AIDS*	*Malaria*	*Reproductive Health*	*Infectious Disease*	*Other Medical*	*Nutrition*	*Other non Medical*	*1st Author Malawian*
1996	26	0	3	2	6	2	2	11	0	0	4
1997	28	2	7	2	1	2	3	10	3	0	3
1998	25	4	5	1	2	5	2	7	0	0	4
1999	25	0	8	0	4	2	3	8	0	0	3
2000	50	2	11	2	7	6	9	12	3	0	2
2001	58	8	10	5	3	2	8	20	1	1	10
2002	49	2	5	10	0	4	12	17	1	0	6
2003	53	2	11	12	2	3	7	15	3	0	9
2004	58	5	11	13	3	2	5	24	2	0	19
2205	54	6	3	9	0	1	3	30	3	0	17
2006	63	4	2	15	7	2	1	28	1	3	25
Totals	489	35	76	71	35	31	55	182	17	4	102
Percentage of total		7.2	15.5	14.5	7.2	6.3	11.2	37.2	3.5	0.8	20.9

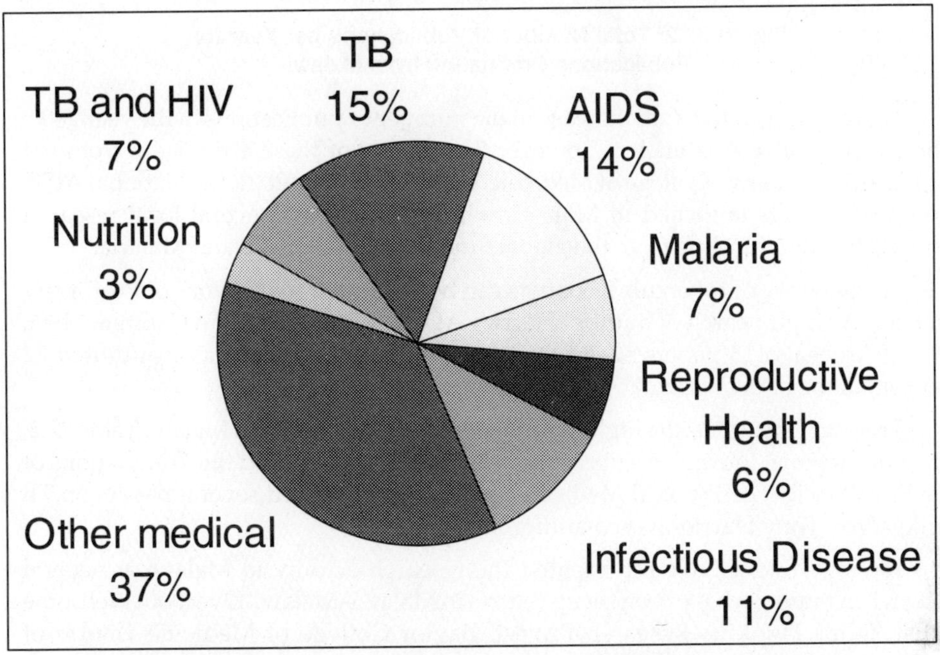

Figure 27.1: Disease Profile of Published Articles Originating from Malawi

Discussion

Our study shows that publications productivity has increased in Malawi by 103 per cent from 26 in 1996 to 63 in 2006 (Figure 27.2). A majority of the papers are published on TB followed closely by HIV and AIDS and then TB and HIV and AIDS jointly. These findings are similar to another study in which for the period 1996-2005 Malawi published 450 papers, this study suggested Malawi had a relative growth of 67.6 per cent (Chatelin and Arvanitis, 1992).

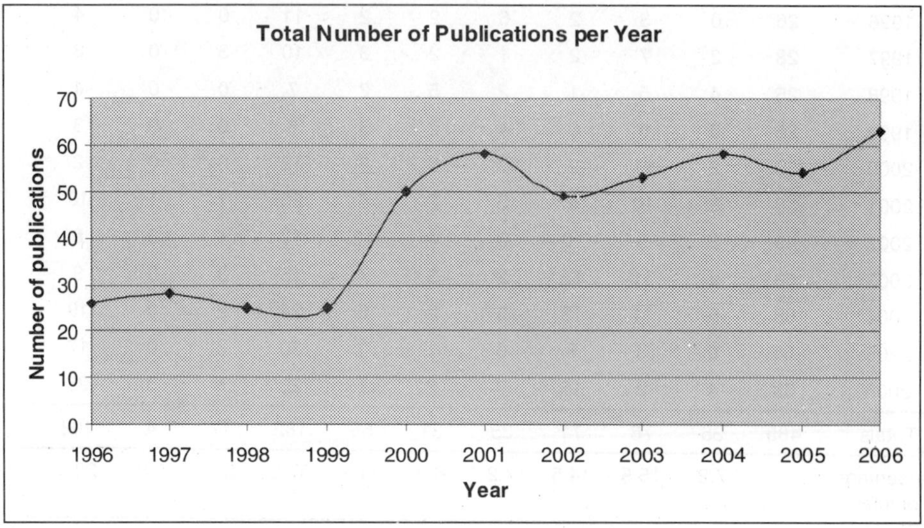

Figure 27.2: Total Number of Publications per Year for Publications Originating from Malawi

There is a rapid two fold increase in the number of publications in the year 2000. The reasons for this, is unclear however 50 per cent of these articles are from the University of Malawi, College of Medicine. In addition from 2001, the National AIDS Commission was launched in Malawi, which has funded several local research projects through funding from the global fund to fight AIDS, TB and Malaria.

The majority of tuberculosis papers can be attributed to the work of Prof. Tony Harries. A single search of author "Harries AD", affiliation "Malawi", limits "1996 to 2006" reveals 113 papers suggesting that Prof. Tony Harries has coauthored 22 per cent of the papers originating from Malawi.

Tropical Doctor has the highest number of publications from Malawi Table 27.2. The International Journal of Tuberculosis and Lung Disease and the Transactions of the Royal Society of Tropical Medicine Hygiene has featured several papers on TB which Prof. Tony Harries has co-authored.

We may under have represented the research activity in Malawi as several research institutions have been set up centres in Malawi–Malawi-Liverpool Wellcome Trust, Johns Hopkins Research Project, Baylor College of Medicine–Centre of

Excellence, University of North Carolina. These international organizations are affiliated to local academic institutions and hospitals and are conducting research on HIV/AIDS, malaria, tuberculosis, nutrition, and maternal mortality. Further investigation is needed to determine what their contribution is to Malawi's scientific output and if it inclusive in the total of 506. Studies examining the number of articles based on research conducted in Malawi and not on affiliation alone may reveal more articles. A comparative study assessing College of Medicine Research and Ethics Committee and Malawi's National Health Research Council approved research proposals to determine research productivity to scientific output may reveal interesting data.

Table 27.2: Top 10 Journals

Title	*Number of Articles*
Trop Doct	34
Inter J Tuberc	30
Trans R Soc	30
East Africa Med.J	25
Lancet	20
Ann Trop Med Parasitol	16
Arch Dis Child	14
AIDS	10
Bull World Health Org	10
J Infect Dis	10

Although the number of papers with Malawian authors is only 20.9 per cent of the total, the number of papers authored has significantly increased from 4 in 1996 to 25 in 2006 which is an over 500 per cent increase Figure 27.3. The number of papers from Malawian researchers was estimated by looking at the names of first authors, this method has limitations. As foreign authors may have Malawian names through marriage and vice versa. Malawian researchers in the Diaspora may be affiliated to non-Malawian institutions and their articles would therefore be excluded from this study. The results exclude Malawians on postgraduate studies who are affiliated overseas but maybe doing research work in Malawi.

The reason of under representation of Malawian first authors may include lack of funding, poor facilities, limited technical support and inadequate training (Soteriades *et al.*, 2006). It may also be the result of Malawian researchers choosing to publish their work in national or regional journals that are not indexed in Medline. These journals are mostly owned by academic institutions, have a small circulations, lack adequate funding, have a difficulty maintaining publication frequency, and are rarely indexed in major databases (Siegfried *et al.*, 2006; Tijssen, 2007). Many researchers also prefer to publish their work in journals with high impact factor and circulation.

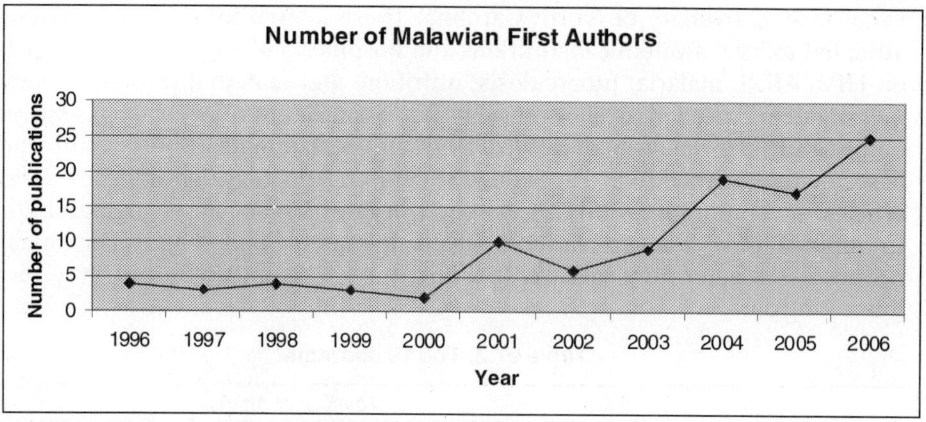

Figure 27.3: Number of Malawian First Authors Publishing Articles per Year

However, these 'western' journals are under some pressure to publish material which is relevant to the majority of the readership and may reject papers from the developing world (Horton, 2003).

Uthman and Uthman's study on Africa biomedical publications ranked Malawi at fifteenth with 450 articles published between 1996 and 2005. South Africa was in first position with 11218 articles. Malawi's relative growth was 67.6 per cent in this study which positioned it at number 13, while Nigeria came top of the list with 2094.7 per cent . The study further breaks down data to population and GDP (Uthman and Uthman, 2007).

The number of articles published per disease type reveals HIV, AIDS and TB as the leading research interest collectively representing 36 per cent which for a country with prevalence of 14 per cent (National AIDS Commission Report, 2003) is commendable but for a country with over a million reported malaria cases in 2000 (Demographic and Health Survey in Malawi, 2000), research on malaria needs to be encouraged. The 37 per cent of other medical research field include papers on ethics and anatomy.

Despite the inherent limitations, our study reveals some general trends in the development of research in Malawi even though the magnitude maybe greater than that found in this present study.

References

Adamson, I, 1992. Access and retrieval of information as coordinates of scientific development and achievement in Nigeria. Scientometrics, 23(1), pp. 191-199.

Badrane, H, Alaoui-el-Azher, M, 2003. Biomedical research in developing countries: the case of Morocco in the 1990s. Tunis Med, 81(6), pp. 377-382.

Chatelin, Y, Arvanitis, R, 1992. Representing scientific activity by structural indicators: The case of Cote d'Ivoire 1884-1968. Scientometrics, 23(1), pp. 235-247.

Demographic and Health Survey in Malawi, 2000

Horton, R, 2003. Medical journals: evidence of bias against the diseases of poverty. Lancet, 361(9359), pp.712-713.

http://www.oecd.org/document/26/0,3343,en_2649_34451_1901082_1_1_1_1,00.html

National AIDS Commission Report, 2003

Perez-Iratxeta, C, Andrade, MA, 2002. Worldwide scientific publishing activity. Science, 297(5581), pp. 519.

Rahman, M, Fukui, T, 2003. Biomedical publication–global profile and trend. Public Health, 117(4), pp. 274-280.

Soteriades, ES, Rosmarakis, ES, Paraschakis, K, Falagas, ME, 2006. Research contribution of different world regions in the top 50 biomedical journals (1995-2002). FASEB J. Jan;20(1), pp.29-34.

Siegfried, N, Busgeeth, K, Certain, E, 2006. Scope and geographical distribution of African medical journals active in 2005. S Afr Med J, 96(6), pp. 533-538.

Tijssen, R, 2007.Africa's contribution to the worldwide research literature: New analytical perspectives, trends, and performance indicators. Scientometrics, 71(2), pp.303-327.

Uthman, OA, Uthman, MB, 2007. Geography of Africa biomedical publications: an analysis of 1996-2005 PubMed papers. Int J Health Geogr. Oct 10, 6, pp. 46.

Chapter 28

Role of COMSATS Science and Technology for Sustainable Development in the South

Tajammul Hussain
COMSATS Headquarters, Islamabad, Pakistan
E-mail: husseint@comsats.net.pk

ABSTRACT

The Commission on Science and Technology for Sustainable Development in the South (COMSATS)–a commission of 21 member countries and 14 Science and Technology centres of excellence across three continents–has been actively engaged in pursuing projects and programmes aimed at assisting member developing countries transit to sustainable paths of development through judicious use of science and technology. Since its inception in 1994 COMSATS has surfaced as an Inter-Governmental Organization (IGO), which is making tangible impacts in the development arena of many member countries through concrete programmes in the realms of Information Communication Technologies, Biotechnology, Renewable Energy Technologies, Climate Change and Environmental Protection, Material Sciences and Water Resources. Key initiatives include the COMSATS Internet Services (CIS), COMSATS Institute of Information Technology (CIIT), Syrian-COMSATS-COMSTECH Information Technology Centre (SCCITC), ICTs for Rural Development in Mountainous and Remote Areas of Northern Pakistan, French Online Distance Learning Programme, the Biotechnology Web-portal and the CERN-COMSATS-NCP Project on Data-Grid Applications and Physics Data-Analysis. COMSATS has firmly placed its faith in the ability of South-South and South-North networking and cooperation to facilitate and harness the future pace of sustainable development in the developing world.

Keywords: *Pakistan, COMSATS, Science and technology, Developing countries, Sustainable development, International agencies.*

Introduction

COMSATS is the brainchild of the Nobel Laureate from Pakistan, Dr. Abdus Salam. The Commission was established in 1994 with the mission of bringing about sustainable socio-economic development in the South through building indigenous capacities in, and making judicious use of science and technology applications, by instituting South-South and North-South collaboration and exchange. Headquartered in Pakistan and having 21 member countries from three continents (Asia, Africa and Latin America), the Commission has remained committed to sensitizing developing nations about the central role of science and technology in improving human conditions, *e.g.* alleviating poverty, improving healthcare, increasing literacy, protecting natural resources and improving economy.

The thrust areas of COMSATS include Information and Communications Technologies (ICTs), Renewable Energy Technologies (RETs), Biotechnology, Climate and Environmental Sciences and development of Small and Medium Enterprises (SMEs). COMSATS is involved in innovative programmes such as sponsorship for scientific activities, postgraduate scholarships, expert-exchange program and distance education. Working partners of COMSATS include many international agencies and organizations such as United Nations Educational, Scientific and Cultural Organization (UNESCO), United Nations Industrial Development Organization (UNIDO), Third World Academy of Sciences (TWAS), Third World Network of Scientific Organizations (TWNSO) and International Atomic Energy Agency (IAEA).

COMSATS is also housing the TWAS/TWNSO regional offices as well as the Pakistan Chapter of ICTP at its Headquarters in Islamabad. COMSATS has been actively implementing various programmes and projects with the aim of assisting its member countries transit to sustainable paths of development. Advocating positive changes in the policy frameworks has been on top of COMSATS' priority tasks. In 2002, COMSATS was accredited as an Inter-Governmental Organization (IGO) with the United Nations Protocol and Liaison Service. The accreditation came as an acknowledgement to COMSATS' effective role in its member countries to bring about sustainable development. In order to further strengthen its roots in the member countries COMSATS is establishing the liaison offices at its focal points. The very first country-liaison office was inaugurated in Sudanese Ministry of Science and Technology in 2004.

COMSATS Network of International Science and Technology Centres for Sustainable Development in the South

One of the main objectives of COMSATS is to support the establishment of a Network of International Science and Technology Centres for Sustainable Development in the South. So far, 15 Centres of Excellence have been included in the Network (14 full members and one associate member) and their services are being channelized to provide the leading role in their respective areas of scientific expertise. This is enabling the other member countries of COMSATS to acquire latest scientific knowledge and skills. These Centres also coordinate with their counterparts in the North for the larger benefit of mankind.

COMSATS International Network of Science and Technology Centres in the South

1. The Biosphere Reserve–Beni, Biology Station (BBS), *Bolivia*
2. National Research Centre for Agrobiology, *Brazil*
3. International Centre of Climate and Environmental Sciences (ICCES), *China*
4. Centro Internacional de Fisica Edificio, *Colombia*
5. National Research Centre (NRC), *Egypt*
6. Building and Road Research Institute, *Ghana*
7. International Centre for Environmental and Nuclear Sciences, *Jamaica*
8. Industrial Chemistry Centre, *Jordan*
9. National Mathematical Centre, *Nigeria*
10. HEJ Research Institute of Chemistry, *Pakistan*
11. Higher Institute for Applied Science and Technology (HIAST), *Syria*
12. Tanzania Industrial Research and Development Organisation, *Tanzania*
13. Marmara Research Centre (MRC), *Turkey*
14. Iranian Research Organization for Science and Technology (IROST), *Iran*

Objectives of the Network

- To assist the countries of the South to build and sustain a critical mass of world-class scientists and technologists in areas of frontier science and technology and environmental sciences which are of critical importance to sustainable socio-economic development through North- South Cooperation.
- To attract talent, reduce brain-drain, and induce competent Third World scientists and technologists working abroad to return to their countries.
- To facilitate the transfer of technology and the supply to industry of highly qualified technical personnel.
- To assist in finding scientific solutions to complex developmental and environmental problems in the South.
- To provide avenues for international cooperation in areas of science and technology of global concern, including those for environmentally sustainable development.
- To provide a suitable framework for science and technology assessment and management in the South.
- To develop strong links between Members of the Network and production sectors in the South.
- To promote joint technological ventures amongst the Members of the Network.
- To assist in carrying out other programmes as assigned by COMSATS.

Functions of the Network

- To conduct high-level training activities, nationally, regionally and internationally, in the scientific fields of each Node of the Network in order to achieve a critical mass of highly-qualified personnel in these fields in the South;
- To conduct basic and applied research in order to develop and test solutions to socio-economic development and environmental challenges facing the South;
- To conduct interdisciplinary research and modelling in order to devise scientific solutions to problems that has global implications for environmentally sustainable development.
- To provide scientific and technical services to the Production and Services sectors in the South.

Invaluable Role of the COMSATS' International S&T Centres of Excellence

To demonstrate the central role played by COMSATS Centres of Excellence in implementing and advancing sustainable development in the countries of the South, case studies of three centres are presented here.

1. National Research Centre (NRC), Cairo, Egypt

Introduction

NRC is the largest multidisciplinary R&D centre in Egypt devoted to basic and applied research within the major fields of interest. It possesses an impressive scientific and technological infrastructure and man power resources of 4300 research staff. NRC consists of 13 divisions and 106 departments covering the major areas of industry, health, environment, agriculture, basic sciences and engineering. It is headed by a president with ministerial status, assisted by two vice presidents, one for research and the other for technical affairs. The minister of state for Scientific Research is the higher president of NRC.

Background

NRC was established as an independent public organization in 1956, with the aim "to foster basic and applied scientific research, particularly in industry, agriculture, public health and other sectors of national economy".

It is the largest of all institutions affiliated to the ministry of Scientific Research and employs about 60 per cent of all scientists working in these institutions. Between the sixties and eighties of the last century six divisions of NRC developed into independent research institutes: The national institute of standards, Petroleum research institute, Central metallurgical research institute, Theodore Bilharz institute, Ophthalmology research institute, Electronic research institute.

Since its establishment, the NRC has passed through three evolutionary stages: initial stage extending to 1968 focused on basic sciences research and capacity building; second stage (1968-1973) was characterized by a growing interaction with the production and service sectors; third stage (1973 till now) concentrating on customer oriented research to serve specific needs of end users.

NRC Vision and Mission

The NRC has to correspond to the country's key production and services sectors through the research conducted in different areas of science and technology, scientific consultation and training as well. The NRC mission is to conduct basic and applied research within the major fields of interest in order to develop production and services sectors.

NRC Objectives

- To provide scientific services and consultations
- To train graduates and researchers
- To contribute in national upgrading of science and dissemination of knowledge
- To strengthen scientific linkages and cooperation with local and international organizations
- To interact with Production and Service Sectors.
- To enhance Advanced Basic Sciences Research.
- To strengthen International Relations.

Interaction with Production and Service Sectors

This plays a vital role in advancing the cause of NRC through establishment of Businessmen and Investors' Office; inclusion of key investors in the Board of directors and divisions; Special Service Units; Oriented Ph.D. research points; Central Unit for Analysis and Scientific Service (CUSS); Training and Capabilities Development Unit; Workshops, symposia, conferences and field visits and studies; Exhibitions for scientific products.

NRC Research Projects

NRC conducts research projects that are customer-oriented while addressing national needs. The in-house projects are done through governmental budget (18 per cent), national projects through contracts (30 per cent) and international projects (52 per cent) through collaboration with foreign partners.

International Publications

NRC has more than 1400 international publications over the years.

Major Areas of NRC-COMSATS Collaboration

NRC Egypt has been very active in its joint programmes with COMSATS and is a major Centre of Excellence in the COMSATS' Network. The mutual ventures include:

- Training
- Conferences and Workshops

- Active participation in COMSATS meetings
- Hosting the 8th and 10th COMSATS coordinating meeting and technical advisory committee meeting
- Visiting professors
- Preparation for collaborative research projects

2. International Centre for Environmental and Nuclear Sciences (ICENS), Jamaica

Introduction

The International Centre for Environmental and Nuclear Sciences (ICENS) is a multi-disciplinary research centre located on the 'Mona Campus' of the University of the West Indies in Jamaica. ICENS is the successor institution to the Centre for Nuclear Sciences (CNS).

The strategic research themes of ICENS mainly involve multi-disciplinary studies that are applicable to a number of environmental and resource components: air, water, soils, rocks, sediments, plants, food animals and humans.

The resulting data, stored in on-line multi-purpose databases and coupled with a Geographic Information System (GIS), are being applied to such areas as agricultural development, environmental assessment and protection, land use, natural resource identification, animal and human health and nutrition. ICENS also assists in the development of human resource to make maximum use of these data.

Mandate and Mission

ICENS' mandate states "It shall be the duty of the Centre to advance science and technology, to foster scientific research in this Island or elsewhere, and to encourage the application of the results of such research for the benefit of the people of Jamaica and collaborating countries."

The mission of the ICENS is to be a scientific institution of the highest quality, which assists in providing solutions for national and regional environmental problems, and plays a role in global issues.

Objectives

The primary objectives of ICENS are:

- To conduct multi-disciplinary research related to the environment
- To help provide solutions to developmental problems
- To contribute to the growth and retention of a cadre of excellent scientists and technologists

ICENS Research and Training Activities

ICENS is conducting basic and applied research in many domains of science and society. The results of this research help solve issues related to Jamaica and its collaborating countries. ICENS is involved in Environmental Geochemical Research in areas that have applications in: Environment; Agriculture; Epidemiology; Nutrition and Health; Occurrence and Effects of Heavy Metals.

- *Agriculture*: Agricultural soils; Impact of trace element geochemistry; Animal and plant health; Hyper accumulators of heavy metals.
- *Human Health*: Human health; Epidemiology.
- *Hydrology*: Detection and minimization of pollution impacts; Likely impacts of drainage and irrigation; Modeling.
- *Mining and Industry*: Mineral exploration; Impacts of mining; Restoration.
- *Geological Mapping*
- *Applications of Isotopes*: Groundwater resources; Fertilizers.
- *Environmental Change*
- *Physical Planning*: Urban and rural planning; National Parks; Tourism; Landslide Risks.
- *Education and Train*ing: Undergraduate studies; Research programmes; Skills training and experience; Public education.
- *Services*: The services provided by ICENS include: X-Ray Fluorescence Spectrometry; Neutron Activation Analysis; Atomic Absorption Spectrophotometry; Ion Chromatography; Radiation monitoring; Personal Dosimetry; Natural Radioactivity.

ICENS Major Scientific Results

In near future, ICENS digital databases and information systems will be steadily enlarged and complemented by others containing geographically referenced information, *e.g.* topography, soil types, climate, vegetation, land use, geology, mineral deposits, plant and animal productivity, selected socio-economic data, health statistics, and medical epidemiology. ICENS has achieved major breakthroughs in its quest of knowledge:

- Numerous publications including a geochemical atlas of 32 elements in Jamaican soils.
- Identification, definition and amelioration of environmental lead hazards.
- Determination of hazardous (heavy metals) substances in soils and foods for policy makers in land use planning and agricultural practices.
- Radon and gold maps.
- Databases which include documents, maps, photographs and satellite imagery. ICENS database contains over 250,000 analytical data entries on over 15,000 samples of soils, rocks, surface and ground water, stream sediments, air particulates, food, blood, animal and human kidney and liver tissues.
- Development of methods for lowering detection limits in neutron activation analysis.

Scientific Achievements/Honors

Researchers from ICENS have received recognition for outstanding work including: National Medal for Science and Technology; Norman Manley Award for Science and Technology; Gleaner Award for Science and Technology.

3. Royal Scientific Society (RSS), Jordan

Introduction

RSS is the largest applied research institution, consultation and technical service provider in Jordan, with more than 600 members of staff. It was established in 1970 as an independent, not-non-profit non-governmental Organization (NGO). RSS is mainly engaged in scientific and technological and Research and Development, consultation, training, and various support services. It is located on a picturesque 85-acre prime site in Al-Jubaiha in northern Amman.

RSS offers both the public and private sectors unique resources and a wide range of project implementation expertise. It enjoys robust IT and communications infrastructures. Most laboratories are nationally and internationally accredited. RSS has close cooperation ties with local, regional, and international organizations. RSS is headed by HRH Princess Sumaya Bint El-Hassan, President RSS assisted by 2 Vice Presidents and Assistant President. There are 13 Departments including University for Technology in Jordan.

RSS Vision and Mission

RSS has the vision to be a leading national S&T institution performing Quality research; Consultations; Technical services; and Specialized training. Its mission is to: serve the Jordanian community to ensure its safety and welfare; promote technological change through research, value-added services, and technology transfer; offer timely solutions, professional expert advice, diversified expertise, manpower enhancement, technical services and support to the various sectors in Jordan and the region.

Aims and Objectives

- To support the development process in Jordan through R&D to strengthen the role of SMEs in Jordan's economy.
- To promote itself as a reference technical institution in Jordan and the region.
- To expand its role as a certification body for both skilled manpower and industrial products.
- To strengthen cooperation with similar institutions to promote mutual interests.
- To develop human resources.
- To encourage and support the start-up, incubation and development of innovation-led knowledge-based businesses.

- CUSS: Central Unit for Analysis and Scientific Service
- CIS: COMSATS Internet Services
- CIIT: COMSATS Institute of Information Technology
- GIS: Geographic Information System
- HIAST: Higher Institute for Applied Science and Technology
- ICENS: International Centre for Environmental and Nuclear Sciences
- IROST: Iranian Research Organization for Science and Technology
- IAEA: International Atomic Energy Agency
- ICTs: Information and Communications Technologies
- ICCES: International Centre of Climate and Environmental Sciences
- IGO: Inter-Governmental Organization
- MRC: Marmara Research Centre
- NGO: Non-Governmental Organization
- NRC: National Research Centre
- PSUT: Princess Sumaya University for Technology
- QAD: Quality Assurance Department
- RSS: Royal Scientific Society
- R&D: Research and Development
- RETs: Renewable Energy Technologies
- SMEs: Small and Medium Enterprises
- SCCITC: Syrian-COMSATS-COMSTECH Information Technology Centre
- S&T: Science and Technology
- TWAS: Third World Academy of Sciences
- TTC: Technology Transfer Centre
- TWNSO: Third World Network of Scientific Organizations
- UNIDO: United Nations Industrial Development Organization
- UNESCO: United Nations Educational, Scientific and Cultural Organization

The Editors

Dr. A.K. Maulloo

Dr. Aman Kumar Maulloo is the Director of the Rajiv Gandhi Science Centre (RGSC) since June 2007. Prior to his present appointment, he has been a Senior Lecturer and Head of the Department of Mathematics, Faculty of Science, at the University of Mauritius. His experience in Operational Research and his affinity for the Sciences are the main ingredients for his leadership at the Rajiv Gandhi Science Centre. As the second Director of this centre, he is continuously striving to make the centre achieve excellence in the promotion and communication of Science and Technology (S&T) among the population. With the introduction of a number of new activities and extending its boundary to reach a wider population band, the image of the centre is continuously improving. Further to the setting up of the new Ministry of Industry, Science and Research in September 2008, the Rajiv Gandhi Science Centre has a more important role than ever before in contributing to the promotion of S&T for the socio-economic development of Mauritius. The dedication and team spirit demonstrated by the members of RGSC and encouraged by Dr Maulloo are the reasons for the high output of the centre. Aman Maulloo has promulgated the centre at the regional level by boosting its relations with its African counterparts by becoming a member of the International Council of African Museums (AFRICOM) and the Southern African Association of Science and Technology Centres (SAASTEC). Further with Mauritius as the present Chair of the South African Development Community Science, Engineering and Technology Week (SADC SET Week), the centre will have an important role to play. International collaboration which is already strong with the National Council of Science Museums (NCSM) India has been extended to France with La Cite des Sciences and de l'Industrie.

Mr. L. Mphahlele

Mr. Lebs Mphahlele has recently been appointed as Director for Curriculum Design in the newly established Academy for Public Administration, Leadership and Management. Between 2004 and 2008 he served as Director for Science and Youth in the newly established Ministry of Science and Technology in South Africa. It was during this period that he played an instrumental role in the development of the Cabinet-approved policy framework for the establishment of a Network of Science Centres in South Africa (see a paper he delivered on this policy in the proceedings of the Hanoi Conference in Vietnam in 2004). He also led the development of a Youth into Science Strategy that will be delivered by the envisaged Network of Science Centres in South Africa. To ensure achievement of the goals of the Network of Science Centres, he facilitated not only the signing of the Collaboration Agreement between the Departments of Science and Technology, and Education in South Africa, but has ensured that this agreement is effectively implemented in all the nine provinces. While employed as both the Chief Education Specialist and Acting Director in the Department of Education, he led, among others, the development of the National Strategy for Mathematics, Science and Technology Education (NSMSTE) and the e-Education White Paper. Some of the achievements of the NSMSTE include increased participation and performance in mathematics and science by the historically disadvantaged. A precursor of the NSMSTE is SYSTEM (Students and Youth into Science, Technology, Engineering and Mathematics) whose curriculum development was managed by Lebs Mphahlele between 1995 and 1998, laid the foundation for post-Apartheid curricula in Mathematics, Science and Technology. More importantly, Lebs Mphahlele played a pioneering role in bringing international conferences on science centres to Africa. These include the 2008 5th International Workshop on Change through Science Centres that took place in Johannesburg and the planned 2011 6th Science Centre World Congress scheduled to take place in Cape Town. He also represented the NAM S&T Centre at the 5th Science Centre World Congress in Toronto, Canada. His contribution and profiling of the role of science centres in developing countries is acknowledged.